D0915924

Assessment in the Mathematics Classroom

1993 Yearbook

Norman L. Webb
1993 Yearbook Editor
University of Wisconsin—Madison

Arthur F. Coxford
General Yearbook Editor
University of Michigan

**National Council of
Teachers of Mathematics**

ISSN 0077-4103
ISBN 0-87353-352-6

Printed in the United States of America

Contents

Part 3: Grades 5–8 Assessment

Part 4: Grades 9–12 Assessment

Part 5: Classroom Assessment Issues to Think About

Thomas J. Cooney, University of Georgia, Athens, Georgia
Elizabeth Badger, Massachusetts State Department of Education, Boston, Massachusetts
Melvin R. Wilson, University of Michigan, Ann Arbor, Michigan

Preface

A yearbook on assessment in the mathematics classroom is not just timely; such a resource is a necessity. As the wave of reform in mathematics education sweeps across many nations, improved mathematics education ultimately depends on the interactions between the teacher and the student. Assessment, the gathering of information for decision making, is critical if one wishes to ensure that those interactions are meaningful learning situations. The "Student Assessment" section of the *Curriculum and Evaluation Standards for School Mathematics* describes the act of teaching as founded on dialogues between teachers and students, each responding to the other on the basis of what has been said or done. Classroom discourse is a main theme in the *Professional Standards for Teaching Mathematics*. Assessment is the effort to decipher the meanings that students assign to the mathematical ideas in these dialogues and to use this information for more effective teaching. Unless there are fundamental changes in the daily interactions that take place between teachers and students, initiatives that would change educational goals, develop new curricula, and devise new methods of instruction will fall short of reaching the reform vision that all students should become mathematically powerful. Assessment is a critical link.

With the advent of the *Standards,* teachers expressed a strong need to know more about assessment. Teachers made it clear that they did not need theoretical discourses on assessment. Rather, they sought examples of new tasks that elicit an integration of mathematical knowledge, new procedures for gathering information on students, and new means to measure over time the progress of students. This yearbook on assessment in the mathematics classroom is grounded in the vision of reform that is expressed in the *Standards*. In addition, it is a response to the many teachers who value change but who need concrete ideas to make that change happen. Consequently, articles for this yearbook were selected using the rigorous criterion that the content had to have clear implications for classroom practices. Whenever possible, students' work and teachers' interpretations are included.

Articles for the yearbook were accepted up until March 1991. The articles can be grouped in three broad areas of interest: techniques of assessment, managing assessment, and issues and perspectives pertinent to classroom assessment. Articles were accepted if they were effective in communicating the view that assessment must be integral to instruction, must support good instructional practices, and must be a multidimensional process that will use a variety of forms. Contributing authors were encouraged to provide specific examples and suggestions for innovative procedures for assessing the overarching curriculum standards of problem solving, communication, reasoning, and mathematical connections within a total assessment plan.

This yearbook is not the final word on classroom assessment; rather, it describes the state of assessment in the mathematics classroom in the early 1990s. Often, articles represent initial efforts to bring assessment into closer alignment with reform goals and objectives. Some of the ideas on assessment that have been advanced most recently, such as the use of portfolios and applications of technology, are mentioned but are not well developed, because they are new to the mathematics classroom and specific illustrations of their use were not available. Given the rapid rate of change in mathematics assessment, it is likely that five years after this yearbook is published, more sophisticated assessment practices will have evolved. One hope for this yearbook is that it will make a positive contribution in directing this change and in influencing assessment practices in the mathematics classroom.

Any yearbook requires the work, time, and energy of many people. Arthur F. Coxford, of the University of Michigan, diligently served as the General Editor. He had the vision for the yearbook, he ensured that timelines were followed, and he contributed greatly in the review of articles. The Advisory Panel consisted of Arthur Coxford; Gail Burrill, of Whitnall High School, Wisconsin; Thomas Romberg, of the University of Wisconsin—Madison; Harold Schoen, of the University of Iowa; and Sharon Senk, of Michigan State University. Their incredible knowledge of mathematics education and assessment was invaluable in reviewing articles, providing productive feedback to authors, and setting high standards for this work. The contributors to this volume should be congratulated for their substantive contributions to assessment. Authors whose articles were not accepted made a contribution by participating in the process. Charles Clements and the NCTM editorial staff went through every article, word by word. The polish that is evident in this yearbook is the result of their efforts.

Participants in this undertaking earnestly hope that the 1993 Yearbook will help all of us to develop our understanding of the essential element in education, specifically, the exchange between teacher and student.

NORMAN L. WEBB
1993 Yearbook Editor

Assessment for the Mathematics Classroom

Norman L. Webb

R APID changes are taking place in conceptions of assessment that focus on the mathematical knowledge of students. A narrow conception of assessment as paper-and-pencil tests that require students to produce an expected—and generally numerical—answer is being challenged by an eclectic view of assessment that draws on many methods to ascertain individual students' knowledge of mathematics. The impetus for this changing view of assessment derives from a strong demand that students know mathematics and be able to use mathematics in the changeable world that these students will face during their lifetimes. The premise of the *Curriculum and Evaluation Standards for School Mathematics* (NCTM 1989) is that students will be using mathematics in a world where calculators, computers, and other forms of technology are readily available; where mathematics as a field of knowledge is rapidly changing; and where mathematics is continually being applied to more fields of work and study.

DEFINITION OF ASSESSMENT

Assessment is the comprehensive accounting of a student's or group of students' knowledge. Assessment is a tool that can be used by a teacher to help students attain the goals of a curriculum. Assessment, and its results, is not—and should not be interpreted as—the end of educational experiences; instead, it is a means to achieve educational goals. Different purposes are served by assessing students' knowledge in the mathematics classroom—measuring students' understanding and use of content, obtaining instructional feedback, grading, and monitoring growth in mathematical achievement. Given these different purposes, teachers are faced with a number of decisions that they must make concerning the form, timing, rigorousness, and usability of assessment.

STANDARDS AND ASSESSMENT

One conception of assessment portrayed in the *Curriculum and Evaluation Standards* treats assessment as a process in which a teacher tries to understand the meanings that students assign to the ideas that are covered in the dialogues that take place between teachers and students during the teaching process. One of the implications of trying to understand the meaning that students assign to ideas is that assessment is treated as though it were a continuous and dynamic process, rather than static testing at specific points in time. The *Standards* make three specific points concerning the ways that assessment can be changed so that it is more adaptable to this dynamic view of assigning meaning to students' ideas and so that assessment enables students to gain the knowledge of mathematics that is envisioned in the *Standards:* student assessment should be integral to instruction; multiple assessment methods should be used; and all aspects of mathematical knowledge and its connections to other branches of knowledge should be assessed.

Conceiving of assessment as a process that is integral to instruction implies approaches to both assessment and instruction. The *Standards* advocates an inquiry-oriented approach to instruction. The statement of standards includes a host of such action words as investigate, extend, develop, explore, relate, conjecture, design, and analyze. Students are to be actively engaged in learning mathematics through activities that are facilitated by the teacher. Problem solving becomes a medium for doing and learning mathematics that pervades instruction rather than a mere expression of outcome. The role of a teacher in an inquiry-oriented instructional situation changes from the traditional instructor as the dispenser of knowledge to the guide or facilitator of instructional experiences. The *Professional Standards for Teaching Mathematics* (NCTM 1991) describes teachers orchestrating oral and written discourse to contribute to students' understanding. Assessment is thus an important tool for understanding the knowledge that students are constructing, the meanings that they are assigning to mathematical ideas, and the progress that they are making toward achieving mathematical power.

Because mathematical thinking is complex and has many aspects, the assessment of this thinking requires the use of different sources of information to ascertain students' development in this thinking. To the extent that the information obtained from multiple assessment sources converges, the teacher gains confidence in the opinions that he or she forms about the mathematical knowledge of students. Any single form of assessment is too limiting, in and of itself, to describe fully a student's knowledge of mathematics.

If assessment practices are to measure students' knowledge of the broad range of mathematical content and the connections among the many ideas and applications of mathematics, in accordance with the goals of the *Curriculum and Evaluation Standards,* then it is necessary to use a variety of assessment methods over time and to consider the different aspects of mathematical knowledge. The *Curriculum and Evaluation Standards* describes a number of curricular topics for each range of grade levels. In addition to more traditional topics, the standards for grades K–4 include patterns and relationships, statistics and probability, and geometry and spatial sense. The standards for grades 5–8 include these topics and algebra. The standards for grades 9–12 include discrete mathematics, the conceptual underpinnings of calculus, and mathematical structure. The *Standards* does not merely enumerate each mathematical topic; it calls for students to be able to make connections among these topics and to relate such connections to situations and applications that are drawn from other disciplines.

FEATURES OF ASSESSMENT

Any form of assessment has five common features. These features apply both to assessment in the classroom and to large-scale assessment. In analyzing any assessment situation, these features provide a framework for discussing the assessment tasks and for reflecting on the form of assessment.

The first feature is the assessment situation, task, or question. The assessment situation could be a problem that is presented to students, a classroom discussion or activity, a question, or any other action that elicits or generates a student's response. One way of creating new forms of assessment is generating new situations.

The second feature is the response. No assessment of a student's knowledge of mathematics is possible without some display of knowledge by the student. A response, however, can take many different forms, such as a written numerical answer, a written paragraph explaining the thinking behind a solution, an oral presentation, a protocol of a student's thinking process that has been obtained in the course of an interview, a journal entry, or a portfolio of a student's work that has been accumulated over time. A variety of forms of responses are necessary to elicit a full range of anyone's knowledge of mathematics.

The third feature is an interpretation of the student's response by a teacher, or by a student in the instance of self-assessment. The interpretation frequently involves comparing the student's response to defined goals or expectations, but it could also involve understanding what the response reveals about the student's structure of mathematical knowledge. An interpretation of a student's response requires making inferences on the basis of an understanding of mathematics, the learning of mathematics, and the knowledge of the student. Any assessment is an inference about what a student knows and can never reveal the full extent or all aspects of that knowledge. More accurate inferences, however, can be made by drawing on information from a variety of sources.

The fourth feature of assessment is assigning some meaning to the interpretation of the student's responses. This meaning could be manifested by locating the student's response on a scale that represents the range of all possible responses. The assigned meaning could be revealed to the student as verbal feedback, indicating how the student was making progress in meeting expectations.

The fifth feature is reporting and recording the findings from the assessment. The reporting of the results of an assessment could be communicated by a written transcript of a student's work or by a simple "Good" written in the margin of a student's work. Generally, the recording of results from an assessment is entered as a grade or a number in a record book. However, some new forms of assessment do not always lend themselves to distillation as numbers but require more extensive reporting and recording procedures. In the course of compiling this yearbook, the authors were strongly encouraged to describe individual assessment techniques by discussing each of the five features of assessment in order to reveal more completely how the techniques are used.

OVERVIEW OF THE YEARBOOK

The yearbook is divided into five sections. The first section includes chapters that discuss some general assessment themes. The next three sections are devoted to explanations of assessment techniques and practices for each of the three grade ranges used in the *Curriculum and Evaluation Standards*—K–4, 5–8, and 9–12. The fifth section treats research and analysis that bears on specific issues in assessment.

General Assessment Themes

Diana Lambdin makes the case in chapter 2 that many of the current recommendations for changes in assessment for the mathematics classroom are not new but have been recommended for a number of years. Donald Chambers emphasizes in chapter 3 the importance of teachers' making instructional decisions on the basis of what is revealed through assessment and discourse about students' thinking. The next two chapters offer perspectives on assessment practices in two other countries—in England in Chapter 4 by Malcolm Swan and in Japan in Chapter 5 by Eizo Nagasaki and Jerry Becker.

Grades K–4

Chapters 6 through 14 present assessment ideas and practices for the primary grades. Chapter 6, by Marja van den Heuvel-Panhuizen and Koeno

Gravemeijer, presents modified written assessment items that have been used in the Netherlands to elicit information about students' thinking strategies. In chapter 7, James St. Clair reports on some observation techniques that he used in a bilingual kindergarten class. Susan Sanford describes in chapter 8 how she was able to turn a show-and-tell experience in her first-grade class into an assessment situation. Guidelines for conducting interviews of elementary students are offered by DeAnn Huinker in chapter 9. Pamela Carter, Pamela Ogle, and Lynn Royer discuss in chapter 10 another technique for collecting information from students—by having students complete learning logs. Chapter 11, by Larry Leutzinger, Myrna Bertheau, and Gary Nanke, describes a classroom in which a teacher has taken a step toward using more alternative forms of assessment by using a variety of forms of assessment in a traditional classroom setting. In chapter 12, Ann Anderson presents some of her findings on efforts to get students to use self-assessment while working with tangrams. The next two chapters detail approaches that have been taken by districts to work with elementary school teachers in an effort to get them to make changes in their assessment practices. Ann Beyer describes in chapter 13 the development of outcome statements and a variety of assessment strategies that can be used to produce information on students' performance. Jane Schielack and Dinah Chancellor, in chapter 14, recount one district's effort to change first-grade assessment practices from multiple-choice tests at the end of a grading period to assessment activities embedded within instruction.

Grades 5–8

Five chapters address specific assessment ideas for the middle grades. In chapter 15, Susan Gay and Margaret Thomas give examples of how students can get the right answer for the wrong reasons or get the wrong answer for the right reason. Linda Dager Wilson and Silvia Chavarria explain in chapter 16 a paper-and-pencil technique called "superitems" and the ways in which this technique can be used to develop profiles of a student's problem-solving abilities. Chapter 17, by Walter Szetela, furnishes examples of how problem statements can be modified or varied to elicit responses from students to assess their critical thinking. The next two chapters, chapter 18 by Kris Warloe and chapter 19 by Maria Mastromatteo, describe assessment activities that are based on the use and analysis of real data by students.

Grades 9–12

A range of assessment techniques are presented in the five chapters that treat assessment in the high school grades. In chapter 20, Denisse Thompson and Sharon Senk discuss algebraic and geometric situations that can be used to assess students' ability to reason and use proofs. Robert Money and Max Stephens explain in chapter 21 four forms of assessment tasks—including investigative projects—that are used in the state of Victoria, Australia. Chapter 22 relates Joan Garfield's use of practical projects in assessing students' understanding of statistical concepts. Jan de Lange reports in chapter 23 on tasks that are included in examinations that students take at the completion of their secondary school experience in the Netherlands. In chapter 24, Lynne Alper, Dan Fendel, Sherry Fraser, and Diane Resek describe some of the ways that the Interactive Mathematics Project has used a combination of assessment procedures, including portfolios.

Assessment Issues

The yearbook concludes with three chapters that discuss specific and important issues in assessment. In chapter 25, Maria Santos, Mark Driscoll, and Diane Briars recount some of their experiences in developing assessment tasks for use in large urban school districts as a part of the Classroom Assessment in Mathematics Network. Assessment tasks based on situations may put some students at a disadvantage because the students may not have the life experiences and belief systems that are required to interpret what is being asked. Patricia Kenney and Edward Silver explore in chapter 26 issues of student self-assessment and provide some examples. In chapter 27, Thomas Cooney, Elizabeth Badger, and Melvin Wilson make a strong case for teachers to relate their conceptions of understanding mathematics to assessment practices. They contend that change in assessment practices should involve more than mere changes in procedures and that change in assessment should include changes in the value that people place on the mathematical knowledge that students can acquire.

The chapters in this yearbook address a range of assessment topics and issues. As a group, they constitute a resource of assessment techniques that can help to fulfill the expectations concerning students' mathematical learning enunciated in the *Curriculum and Evaluation Standards for School Mathematics* and the *Professional Standards for Teaching Mathematics.*

REFERENCES

National Council of Teachers of Mathematics. *Curriculum and Evaluation Standards for School Mathematics.* Reston, Va.: The Council, 1989.

_____. *Professional Standards for Teaching Mathematics.* Reston, Va.: The Council, 1991.

2

The NCTM's 1989 Evaluation Standards: Recycled Ideas Whose Time Has Come?

Diana V. Lambdin

WHEN the National Council of Teachers of Mathematics (NCTM) published its *Curriculum and Evaluation Standards for School Mathematics* in 1989, many readers may have felt that its recommendations seemed novel. For example, the *Standards* recommended assessing domains that seemed unfamiliar to many teachers: "disposition toward mathematics" (p. 205), ability to "translate from one mode of representation to another" (p. 223), and ability to "express mathematical ideas by speaking, writing, demonstrating, and depicting them visually" (p. 214). Also, many of the recommended assessment methods were different from those routinely used in mathematics classrooms of the 1980s: for example, having students write essays about their understanding of mathematical ideas, and using classroom observations and individual student interviews as methods of assessment. Although some of the standards may have seemed innovative, their authors never claimed that they were a new revelation. In fact, the *Standards* document is a compilation of recommendations, many of which have been discussed for decades but all of which seem especially appropriate, in recycled form, for today's classrooms and for those of the foreseeable future.

Take a moment to examine figure 2.1. Each of the five quotations in the figure advocates assessment techniques recommended in the *Standards,* yet each was written in a different decade from the 1940s to the 1980s. Only one of the recommendations is actually taken from the *Standards.* See if you can identify the correct decade for each quotation. (Dates and references are provided at the end of the chapter.)

Perhaps you had little difficulty separating the older quotations in figure 2.1 from the more recent ones: in several cases, the language tends to give it away. However, if you look below the surface features of the statements, the similarity

7

1. In general, observation, discussion, and interview serve better than paper-and-pencil tests in evaluating a pupil's ability to understand the principles and procedures he uses.

2. Information is best collected through informal observation of students as they participate in class discussions, attempt to solve problems, and work on various assignments individually or in groups.

3. Evaluation of the thinking and procedures employed by students usually is better done by careful observation and interview than by objective testing.

4. From the standpoint of the classroom teacher in particular, frequent *informal* observations of student behavior have a vital role to play in the evaluation process. They neither replace nor are replaced by the more formal observations of student behavior that are made on the basis of tests and the like. (Emphasis in original)

5. Observation of the pupil's oral and written work … [is] a very important testing procedure and should be encouraged. Closely associated with the observation technique is the interview with the pupil regarding his daily work or his solution or attempted solution of items of a test.

Fig. 2.1. Recommendations concerning evaluation, authored from 1946 to 1989

of their messages is evident. All five quotations, drawn from documents published in five different decades over the past half-century, advocate increased use of informal, observational methods of assessment. These statements provide just one illustration that the NCTM *Curriculum and Evaluation Standards* is trumpeting a collection of recommendations that actually are not new.

ECHOES FROM THE PAST

Let us look more closely at three examples of how the *Standards* echoes ideas from the past. The examples discussed below involve mathematical concepts, mathematical power and disposition, and multiple sources of assessment information.

Mathematical Concepts

Much recent discussion about the goals of mathematics learning has focused on the development of the understanding of mathematical concepts (National Research Council 1989; NCTM 1991). "Concepts are the substance of mathematical knowledge," according to Evaluation Standard 8—Mathematical Concepts (NCTM 1989, p. 223). Because conceptual knowledge develops slowly, fundamental ideas that are introduced in the early grades are repeatedly elaborated on and extended as students move through school. Understanding concepts "involves more than mere recall of definitions and recognition of common examples.... Tasks that ask students to apply information about a given concept in novel situations provide strong evidence of their knowledge and understanding of that concept" (p. 223). Authors in the 1940s used different language to make a similar point about the importance of assessing conceptual understanding by using novel situations:

> In preparing test items, try to modify the conventional forms of presenting exercise material so that rote learning is less likely to provide successful responses and that more thorough understanding of principles is rewarded. (Hartung and Fawcett 1946, p. 163)

Mathematical Power and Mathematical Disposition

Although conceptual understanding is clearly one of the major goals of mathematics teaching, students' capacity for integrating, applying, and communicating their mathematical understandings is also important. Evaluation Standard 4 refers to this capacity as *mathematical power*, asserting that "the assessment of students' mathematical power goes beyond measuring how much information they possess to include the extent of their ability and willingness to use, apply, and communicate that information" (p. 205). Furthermore, Evaluation Standard 10—Mathematical Disposition maintains that it is also important to assess such things as students' confidence, interest, curiosity, and inventiveness in working with mathematical ideas. Corcoran and Gibb (1961) and other writers in the 1950s and the 1960s argued similar points:

> One of the best indications of the mastery of a subject possessed by a pupil is his ability to make significant comments or to ask intelligent questions about the subject.... Another indication of achievement in a field is interest in that field.... Still another indication of achievement is the degree of confidence displayed when work is assigned or undertaken. (Spitzer 1951, pp. 193–94)

> Appraisal ideally includes many aspects of learning in addition to acquisition of facts and skills. It includes the student's attitude toward the work; the nature of his curiosity about and ingenuity with mathematics; his work habits and his

methods of recording steps toward a conclusion; his ability to think, to exclude extraneous data, and to formulate a tentative procedure; his techniques and operations; and finally, his feeling of security with his answer or conclusion. (Sueltz 1961, pp. 15–16)

Multiple Sources of Assessment Information

If students demonstrate their mathematical power through a multiplicity of behaviors—being able to ask important questions, to make connections, to apply understandings to novel situations, and to do so confidently and efficiently—then teachers must also use a variety of methods for assessment. Thus, another recurring theme in the *Curriculum and Evaluation Standards* is the need for Multiple Sources of Information—Evaluation Standard 2. The *Standards* recommends the use of diverse assessment techniques:

> Observing students solving problems individually, in small groups, or in whole-class discussions; listening to students discuss their problem-solving processes; and analyzing tests, homework, journals, and essays. (P. 209)

> When teachers find that students perform in consistent ways on . . . tasks that demand a range of mathematical thinking or represent different aspects of mathematical thought, teachers can have confidence in the accuracy of their judgments. (P. 196)

Yet the recommendation that teachers use multiple sources of information for assessment is hardly new. Brownell (1946), a prominent mathematics educator of the 1940s, wrote:

> Evidences of learning abound on every hand, provided that teachers are alert to their presence and to their significance. Some of this evidence is susceptible to measurement by means of paper-and-pencil tests. Other evidences of learning are best assessed in other ways, for example, by examining pupils' work products, by questioning pupils in the classroom and in conferences, and by observing their behavior in and out of school. Such opportunities to evaluate learning are too important to be neglected. (P. 1)

Findley and Scates (1946) and Hartung (1961) expressed similar views. Brownell's statement, which sounds contemporary even today, comes from his introduction to *The Measurement of Understanding,* a 1946 yearbook of the National Society for the Study of Education (NSSE). In the concluding chapter of that yearbook is something even more intriguing. That chapter makes it clear that educators in previous decades recognized the discouraging tendency for similar recommendations to be repeated year after year with little change in classroom practice. Not only were the editors of the 1946 NSSE yearbook aware of the frequency—and the ineffectiveness to date—of exhortations about the importance of teaching for mathematical understanding, but they also recognized that appropriate assessment techniques must be the driving force in promoting such teaching:

It is highly probable that for several decades every national committee on the teaching of the various school subjects has included in its list of objectives statements which relate to understandings.... In a word, educational theory has consistently given prominence to the necessity of cultivating understandings through instruction. The same statement cannot be made, however, with regard to practice in classroom instruction.... [I]t is a fact that in our teaching we are still prone to pay lip-service to understandings as educational aims.... By the same token, we shall have to commit ourselves seriously to the evaluation of understandings.... When we begin systematically to assess understandings, we must be prepared to alter some of our ideas about evaluation. (Brownell et al. 1946, pp. 321–22)

Why have many of the same ideas about how to teach and how to assess been recommended for more than half a century but never successfully implemented? More important, why should we now—in the final decade of the century—concern ourselves with the recommendations of the *Standards?* The pages that follow speculate about why past recommendations have not been heeded and discuss reasons why such recommendations may still be difficult to implement today.

WHAT HAS IMPEDED IMPLEMENTATION IN THE PAST?

Until quite recently, despite clear and consistent recommendations, few teachers have concerned themselves with assessing students' abilities to make mathematical connections or communicate mathematically. Why? Certainly, teachers themselves are not to blame. The most compelling answer may be that American society as a whole, and much of the educational community as well, has persisted since the early days of this country in viewing mathematics as a school subject that developed and exercised the reasoning powers of the mind but whose specific content was restricted to number (and perhaps shape) and the ability to manipulate these entities according to prespecified rules (Cohen 1982). Thus, in spite of impassioned arguments through the years about broader goals for mathematics teaching, teachers—in response to society's demands—have primarily concentrated on computational facility, on the mastery of discrete skills, and on the ability to solve problems similar to those presented in the textbook or in class. Reinforcement for these emphases came from the fact that standardized tests rarely included items designed to assess higher-order thinking. As long as society—embodied, in particular, by standardized tests—focused attention on lower-level mathematical skills, these skills continued to be the priorities of classroom teachers and the objects of their classroom assessment efforts. In the language of the *Standards,* teachers have been consistent in *aligning* their assessment goals with their teaching goals and with the goals of society. External assessment expectations seem to have a profound influence on what is valued in the classroom (cf. Romberg, Zarinnia, and Williams 1989).

Until quite recently, conventional wisdom assumed that traditional tests provided adequate measures of students' mathematical abilities. Students were judged to have "done well" in mathematics if they were prepared for work in subsequent classes and achieved high scores on tests imposed by districts, states, and external agencies. Even twenty years ago, no matter how much mathematics educators may have called for teaching and assessing mathematical understandings, it was still commonly believed that one of the most important skills required of high school graduates was the ability to calculate and manipulate symbols quickly and accurately. Consequently, these were the skills emphasized on many tests.

WHAT PROBLEMS WITH ALTERNATIVE ASSESSMENT PERSIST TODAY?

There is no doubt that new assessment techniques continue, even today, to present teachers with difficulties that are not easily overcome. It seems only fair to acknowledge some of these problems. First of all, nontraditional assessment methods are often more difficult to design and implement. In particular, much of the assessment data recommended by the *Standards* is more difficult to collect than test scores. Classroom observations are a case in point: it takes considerable expertise to circulate around a classroom as students work in small groups, simultaneously providing assistance and guidance and also making mental notes of students' strengths, weaknesses, and dispositions. Individual student interviews are another example: in a normal classroom situation, time pressure and concern for classroom discipline often interfere with opportunities for extended talk with individuals.

Another difficulty is that, once collected, nontest data are more difficult to organize, summarize, and report. Test scores in a gradebook take up little space and are easily combined to produce a single grade. Notes from observations or interviews and written comments on student essays or portfolios are more cumbersome to maintain and more difficult to sift and mold into a concise, focused evaluation.

Furthermore, work with nontest data is almost always more time consuming, especially when a teacher has many students. Elementary school teachers who work with just one class of students may find assessing student portfolios a reasonable task. But even carrying the portfolios home to read is no simple matter for a high school teacher who may have as many as five or six classes of twenty-five to forty students each.

Finally, the most formidable impediment to innovative assessment techniques may be tradition. Educational assessment procedures that have been in place for decades are difficult to change. Tests and letter grades are well established as methods for evaluating and reporting students' achievements in mathematics. Because Americans have been fascinated ever since the end of the eighteenth century with measuring and numerically describing all types of national trends (Cohen 1982), it is no surprise that the society as a whole seems fixated on the importance of numeric evidence of student progress (witness the hue and cry each year when SAT scores for the fifty states are publicized). Even if teachers are convinced of the benefits of using more innovative methods to evaluate their students, they are unlikely to succeed unless their supervisors, students, parents—and even their fellow teachers—understand and support their break with tradition.

WHY SHOULD TODAY BE DIFFERENT?

Ideas similar to those of the *Standards* have been recommended for decades, yet rarely implemented. Many of the ideas are not only unfamiliar but also—admittedly—more difficult to implement. Why, then, should today be any different? There are several reasons for optimism.

To begin, there is the well-recognized national furor over the inability of students to think for themselves, to solve problems, to demonstrate number sense, and to reason creatively (National Research Council 1989). Although teachers are not surprised by such claims, the furor may bring long-ignored goals to the attention of the public and education authorities (Blackwell and Henkin 1989; Mathematical Sciences Education Board 1990). Perhaps just as important, education officials, realizing the driving power of standardized tests, are beginning to modify the form of many of these instruments. It is only to the extent that estimation, open-ended problem solving, and written expression are valued by those in authority—and are included in external assessments—that such abilities will receive greater emphasis in curriculum, instruction, and classroom assessment.

A second reason for optimism is that technological advances have introduced the possibility that alternative assessment methods will no longer be so cumbersome to implement. Computers, graphing calculators, videocassette and videodisk recorders, and camcorders can be used in innovative ways for both instruction and assessment. Observation in the classroom becomes more feasible when one can record and later play back the work of a small group for analysis. Data management and word-processing programs may make it almost as efficient to write up and organize notes about students as it used to be to enter numeric grades in a gradebook. Certainly, it is much easier to create test situations in which students deal with realistic numbers when computers can simulate real-world situations or when every student can have a calculator readily at hand.

Still another reason for optimism today is that the *Standards* documents enjoy much more broad-based support than many of the recommendations of decades past. The committees who drafted the *Standards* were composed of representatives from a carefully chosen mixture of constituencies:

mathematicians, educators, administrators, policy analysts, and—most important—classroom teachers. As a result, the *Standards* documents have had much more influence than any journal article, yearbook chapter, or association press release could ever hope to have. Furthermore, the *Standards* represents a professional consensus that can provide political support and moral suasion for those who try to implement the changes that the documents suggest.

Finally, and perhaps most important, we may be able to be optimistic concerning change this time because so many *teachers* have ownership in the ideas proposed in the *Standards*. History shows that changes ordained from above are almost certainly doomed to failure, whereas grass-roots efforts stand a chance—as long as there is a large enough groundswell to support them. Classroom teachers were involved from the very beginning in the *Standards*: drafting and critiquing the documents; giving workshops to explain them to parents, administrators, and peers; and writing supplementary materials (such as the booklets in the NCTM's Addenda Series) to provide concrete ideas for how to implement the *Standards* in the classroom.

NEXT STEPS

The heading for this final section, "Next Steps," echoes three earlier documents: "Next Steps" is not only the title of the final chapter of the 1946 NSSE yearbook but also the heading used for the final section of the 1989 *Curriculum and Evaluation Standards* as well as the final section of the 1991 *Professional Teaching Standards*. What are the next steps?

If classroom practices are already changing, then the logical next step is for classroom assessment to change as well. In fact, this is the message of the first of the NCTM's evaluation standards—Alignment: "Methods and tasks for assessing students' learning should be aligned with the curriculum's goals, objectives, and mathematical content" and with its "instructional approaches and activities" (NCTM 1989, p. 193). Teachers might begin with the following five "next steps" (Lester and Kroll 1991):

Step 1: Start small

Step 2: Incorporate assessment into the class routine

Step 3: Set up an easy and efficient record-keeping system

Step 4: Establish an assessment plan

Step 5: Personalize the assessment plan

WHAT HAVE WE LEARNED?

Perhaps the most valuable lesson we can learn from a historical look at writings about mathematics assessment is that productive change is not inevitable. Task forces may produce recommendations, national leaders may give speeches, and educational organizations may publish yearbooks, but progress will be only incremental at best unless all of us who teach assume ownership of new ideas and work to see that they are implemented in our classrooms.

But we cannot implement changes alone. And it is precisely on this point that optimism about productive change must hinge. Even though ideas that have been recommended for years are being touted once again in the 1990s, there is a new sense of interest and commitment in the air. At the national level, standardized tests are being modified to include attention to higher-order thinking; at the state level, new assessment instruments are being adopted and new priorities endorsed. As a result, we can be more optimistic than ever about the prospects for productive change in our nation's classrooms as well. Alternative assessment techniques, such as those recommended by the NCTM *Standards* (1989), seem to be recycled ideas whose time has finally come.

Dates and References for Quotes in Figure 2.1

1. Sueltz, Boynton, and Sauble (1946, p. 145)
2. NCTM (1989, p. 233)
3. Sueltz (1961, p. 17)
4. Weaver (1970, p. 336)
5. Spitzer (1951, p. 191)

REFERENCES

Blackwell, David, and Leon Henkin. *Mathematics.* Report of the Project 2061 Phase I Mathematics Panel. Washington, D.C.: American Association for the Advancement of Science, 1989.

Brownell, William A. "Introduction: Purpose and Scope of the Yearbook." In *The Measurement of Understanding,* Forty-fifth Yearbook of the National Society for the Study of Education, Pt. 1. Chicago: University of Chicago Press, 1946.

Brownell, William A., Harl R. Douglass, Warren G. Findley, Verner J. M. Sims, and Herbert F. Spitzer. "Next Steps." In *The Measurement of Understanding,* Forty-fifth Yearbook of the National Society for the Study of Education, Pt. 1. Chicago: University of Chicago Press, 1946.

Cohen, Patricia Cline. *A Calculating People: The Spread of Numeracy in Early America.* Chicago: University of Chicago Press, 1982.

Corcoran, Mary, and E. Glenadine Gibb. "Appraising Attitudes in the Learning of Mathematics." In *Evaluation in Mathematics,* Twenty-sixth Yearbook of the National Council of Teachers of Mathematics, edited by Donovan A. Johnson, pp. 105–22. Washington, D.C.: The Council, 1961.

Findley, Warren G., and Douglas E. Scates. "Obtaining Evidence of Understanding." In *The Measurement of Understanding,* Forty-fifth Yearbook of the National Society for the Study of Education, Pt. 1. Chicago: University of Chicago Press, 1946.

Hartung, Maurice L. "Basic Principles of Evaluation." In *Evaluation in Mathematics,* Twenty-sixth Yearbook of the National Council of Teachers of Mathematics, edited by Donovan A. Johnson, pp. 21–42. Washington, D.C.: The Council, 1961.

Hartung, Maurice L., and Harold P. Fawcett. "The Measurement of Understanding in Secondary-School Mathematics." In *The Measurement of Understanding,* Forty-fifth Yearbook of the National Society for the Study of Education, Pt. 1. Chicago: University of Chicago Press, 1946.

Lester, Frank K., Jr., and Diana Lambdin Kroll. "Evaluation: A New Vision." *Mathematics Teacher* 84 (April 1991): 276–83.

Mathematical Sciences Education Board. *Reshaping School Mathematics: A Philosophy and Framework for Curriculum.* Washington, D.C.: National Academy Press, 1990.

National Council of Teachers of Mathematics. *Curriculum and Evaluation Standards for School Mathematics.* Reston, Va.: The Council, 1989.

_____. *Professional Standards for Teaching Mathematics.* Reston, Va.: The Council, 1991.

National Research Council. *Everybody Counts: A Report to the Nation on the Future of Mathematics Education.* Washington, D.C.: National Academy Press, 1989.

Romberg, Thomas, E. Ann Zarinnia, and Steven R. Williams. *The Influence of Mandated Testing on Mathematics Instruction: Grade 8 Teachers' Perceptions.* Madison: National Center for Research in Mathematical Sciences Education, University of Wisconsin, 1989.

Spitzer, Herbert F. "Testing Instruments and Practices in Relation to Present Concepts of Teaching Arithmetic." In *The Teaching of Arithmetic,* Fiftieth Yearbook of the National Society for the Study of Education, Pt. 2, edited by Nelson B. Henry, pp. 186–202. Chicago: University of Chicago Press, 1951.

Sueltz, Ben A. "The Role of Evaluation in the Classroom." In *Evaluation in Mathematics,* Twenty-sixth Yearbook of the National Council of Teachers of Mathematics, edited by Donovan A. Johnson, pp. 7–20. Washington, D.C.: The Council, 1961.

Sueltz, Ben A., Holmes Boynton, and Irene Sauble. "The Measurement of Understanding in Elementary-School Mathematics." In *The Measurement of Understanding,* Forty-fifth Yearbook of the National Society for the Study of Education, Pt. 1. Chicago: University of Chicago Press, 1946.

Weaver, J. Fred. "Evaluation and the Classroom Teacher." In *Mathematics Education,* Sixty-ninth Yearbook of the National Society for the Study of Education, Pt. 1, edited by E. G. Begle. Chicago: University of Chicago Press, 1970.

3

Integrating Assessment and Instruction

Donald L. Chambers

THE *Professional Standards for Teaching Mathematics* (National Council of Teachers of Mathematics 1991) describes teaching as a complex interaction of—

- selecting, developing, or modifying appropriate instructional tasks;
- orchestrating classroom discourse;
- establishing and maintaining a stimulating classroom environment; and
- analyzing the impact of the tasks, discourse, and environment on student understanding.

Assessment, according to Webb and Briars (1990, p. 108),

> must be an interaction between teacher and students, with the teacher continually seeking to understand what a student can do and how a student is able to do it and then using this information to guide instruction.

This view is consistent with the Evaluation Standards proposal that "student assessment be integral to instruction" (NCTM 1989, p. 190). Every instructional activity is an assessment opportunity for the teacher as well as a learning opportunity for the student.

> Well-chosen tasks afford teachers opportunities to learn about their students' understandings even as the tasks also press the students forward. (NCTM 1991, p. 27)

This article is on the interaction between the selection of tasks and the analysis of each student's response to the tasks.

The author is grateful to Elizabeth Fennema and Norman Webb for their contributions to the preparation of this article.

17

COMPONENTS OF TEACHING

Teachers can use task selection, discourse, and analysis as foci for thinking about both instruction and assessment. These terms are defined by the *Professional Standards for Teaching Mathematics* (NCTM 1991, p. 20) as follows:

- *Tasks* are the projects, questions, problems, constructions, applications, and exercises in which students engage.

- *Discourse* refers to the ways of representing, thinking, talking, and agreeing and disagreeing that teachers and students use to engage in those tasks.

- *Analysis* is the systematic reflection in which teachers engage. It entails the ongoing monitoring of classroom life—how well the tasks, discourse, and environment foster the development of every student's mathematical literacy and power. Through this process, teachers examine relationships between what they and their students are doing and what students are learning.

These components are as helpful on the first day of school, when the teacher knows little about each student's understanding, as they are on a day in May, when the teacher knows about each student's understanding in considerable detail. All teachers can learn how to acquire detailed knowledge of each student's mathematical thinking and understanding and how to use that knowledge as the basis for instructional decisions both for individual students and for the class as a whole.

TEACHERS' KNOWLEDGE

Children solve problems in well-identified ways. Teachers who have a structured knowledge of both the mathematical domain under investigation and the nature of children's thinking within that domain can more readily and accurately assess a child's thinking within that domain. A teacher should be aware, for example, of the difference between a measurement division problem such as

José has 15 toy trucks. If each friend gets to play with 3 trucks, how many friends can José have to play?

and a partition division problem such as

José has 15 toy trucks. If he lets 3 friends each play with the same number of his trucks, how many trucks can each friend play with?

Young children tend to model the semantic structure of the problem rather than its mathematical structure. In the first problem a child may form groups of three from a collection of fifteen counters to arrive at an answer of five; five groups of three trucks. The second problem features three groups, not groups of three. Here a child might guess an answer, say four, and then form groups of four. The child would then count the groups to see if three groups of four could be formed, and adjust the guess if it didn't work out. Another strategy for this partition division problem would be to deal out the trucks (or counters representing trucks) to three real or imagined friends. This would produce three groups of five trucks.

Some children might recognize that the two problems have the same mathematical structure (What number of threes is fifteen?) and use a measurement strategy for a partition problem. They might remove a group of three, thinking of it as one truck for each of the three friends. A second group of three is a second truck for each friend. They would continue removing groups of three, as in the measurement division problem, until all possible groups of three had been removed.

In general, partition division problems are harder for young children to solve than measurement division problems. Children who use such a measurement strategy for a partition problem are usually more advanced in their thinking than students who use only partition strategies. Teachers who are unaware of these distinctions are more limited in their ability to assess students' thinking and to make instructional decisions based on that assessment. More knowledgeable teachers will not simply say that a student can or cannot solve division problems. The teacher will be able to identify which students can solve each of the two division types and also what solution strategy each student uses for the types he or she can solve. Bebout and Carpenter (1989) furnish examples of a teacher assessing students' understanding using a structured framework of problem types and solution strategies.

Sammons, Kobett, Heiss, and Fennell (1992) point out that specific training is necessary for teachers to learn to assess children's thinking by analyzing student discourse. Carpenter, Fennema, Peterson, Chiang, and Loef (1989) have described one approach to helping teachers learn to assess children's thinking. They help teachers learn to structure their knowledge of the content domain by identifying problem types and the solution strategies students might use.

The problem types tend to form a hierarchy, with some types being more difficult for children to solve than others. Generally children are able to solve problems of the easiest type first and systematically become successful with more and more difficult types as they acquire more and more mathematical experience. The distinction between measurement division problems and partition division problems mentioned earlier is an example of this hierarchy.

Children's solution strategies can also be identified and organized into a hierarchy. The earliest strategies to emerge are direct-modeling strategies, in which the children represent the objects in the problem by counters and manipulate the counters in the fashion indicated by the problem statement. A student using a direct-modeling strategy for the measurement division problem above would count out fifteen counters and then remove the counters three at a time until all fifteen counters had been arranged into groups of three. The student would then count the groups to determine the answer.

A student using a somewhat more advanced strategy, a counting strategy, might solve the same problem by counting to 15 and raising a finger after each third number, as follows: one, two, three (raise a finger); four, five, six (raise a finger); seven, eight, nine (raise a finger); ten, eleven, twelve (raise a finger); thirteen, fourteen, fifteen (raise a finger). The student would then count the number of fingers raised to determine the answer. Counting is used in both strategies. The strategy referred to as a counting strategy differs from the direct-modeling strategy in that the initial number, 15, is not represented by fifteen objects. The physical objects used, fingers in this situation, are used to keep track of counts rather than to represent objects in the problem.

Even more advanced strategies are number-facts strategies. In the measurement division problem the student may say, "I knew that 3 times 5 is 15" or "I knew that 15 divided by 3 is 5." Students may know some facts but not the one that solves the problem. When they use a known fact to obtain an unknown fact, the students are using derived facts. An example of a derived-fact strategy for this problem is "I knew that 3 times 10 is 30, so 3 times 5 is 15." This child realizes that 15 is half of 30, so a multiplier that is half of 10 is needed.

COMPONENTS OF ASSESSMENT

Getting the students to understand how they, the teachers, think about mathematics seems to be the goal of many teachers. More consistent with the vision of the *Standards* is getting the teacher to understand how the students think about mathematics. To assess students' understanding during instruction, teachers need to collect data on student understanding by observing and listening to student discourse and interpret the data to develop an accurate description of student thinking. Then, using their knowledge of the thinking of each student, teachers can select appropriate instructional tasks.

Teachers can learn to orchestrate discourse to elicit information that reveals student thinking. Questions should focus on the student's solution strategy rather than on the answer. Such questions as

"How did you solve this problem?"
"Did anyone else use the same strategy?"
"Did anyone use a different strategy?"
"Can anyone think of another strategy?"
"Tom, what do you think about Ellen's strategy?"

are designed to reveal how students think about the problem. Evidence of student thinking may also come from their written work and from any manipulation of objects that accompanies their solutions. When the evidence is inadequate, a teacher may have to ask probing questions.

The following vignette illustrates how teachers can use the results of informal assessment to help them make informed instructional decisions.

Ms. Grainger's Class

It is the twentieth day of school, and Ms. Grainger has already learned a great deal about the mathematical thinking of each of her twenty-three first-grade students. She knows what size of numbers each student can use to solve problems. The students have solved a variety of problems that involve joining and separating. Ms. Grainger can talk with some confidence about the strategies that different children use to solve problems. Most of the children use direct modeling, that is, they heed the action described in the problem (joining, separating, etc.) and use counters (objects, fingers, pictures, tally marks) to represent the initial situation, the change, and the final situation. Because Ms. Grainger has carefully selected problems that are interesting to the children and that they can solve using the informal knowledge they already possess, the children have confidence in their ability to solve problems.

Ms. Grainger has not explained to the students how to solve story problems, nor has she explained how to add and subtract one- and two-digit numbers. Ms. Grainger poses a problem she knows will be difficult for some of the students:

Tomorrow I will be taking treats for my son's soccer team. I will take Kit Kat bars. I have 19 Kit Kat bars already. There are 27 players on my son's soccer team. How many more Kit Kat bars do I need?

Ms. Grainger knows that many children will need a lot of time to solve this problem, and she knows who they will be. She knows that Lisa and Martin will get the solution fairly rapidly because they use derived facts and do not have to model the problem. Lisa and Martin will need additional tasks while the other children continue to work on this one. If Lisa and Martin respond as she expects, she will pose a similar problem with a gap between the numbers that is greater than 10.

She knows she may need to modify the problem for Anna or Bruce, who may have trouble working with numbers this large. For them she may restate the problem with smaller numbers, or she may use numbers of approximately the same size but use two numbers from the same decade or make the gap between the two numbers smaller.

Because the problem statement does not suggest separating, Ms. Grainger does not expect that any students will solve the problem by subtracting 19 from 27.

Ms. Grainger watches the students as they work and listens to them as they talk. She notices that Norman is not engaged, but she knows that Norman does not need help in understanding the problem and so does not provide any help. She stands near Norman to indicate to him that it is time to work.

Ms. Grainger sees that Bruce is having trouble getting started. Anna seems to be working productively counting out 19 colored chips, but Bruce doesn't seem to know how to begin. Ms. Grainger modifies the problem for Bruce, telling him she has 8 Kit Kat bars and wants to have 11. Bruce counts out 8 counters and then continues counting, "Nine, ten, eleven." Bruce was able to determine that Ms. Grainger needed 3 more Kit Kat bars. Ms. Grainger then repeated the original problem for Bruce.

Martin and Lisa have finished and Ms. Grainger asks them to explain their solutions to her. Lisa notes that if Ms. Grainger had had 20 Kit Kat bars, she would need 7 more. Since she had only 19, she would need 8 more. Martin knew that 9 and 9 is 18, so 9 and 19 is 28. To get 27, she needed 1 less than 9, or 8. Ms. Grainger asks them how many she would need if she had 25 and she wanted to have 58. Without waiting she goes to check on Tom.

Tom has counted out 19 Unifix cubes. He then continues counting until he has 27 cubes. Ms. Grainger notes that Tom was unable to plan ahead to realize he had to keep the counters for 20 through 27 separate from the others so he could go back and recount them. She knows this feature of this problem makes it more difficult than a problem in which the students combine two separate amounts to get the total. She knows Tom can do that. She wonders what question she can ask that will help Tom think about the problem. She asks him to show her which bars he had to start with. She watches as Tom counts out the original 19 from the group of 27. When she asks how many more were bought, Tom counts the remaining cubes and responds, "Eight." Ms. Grainger makes a brief note about Tom's need for help.

For taking notes, Ms. Grainger carries a clipboard with a strip of self-adhesive mailing labels. She notes the student's name and date and a brief note about the student's thinking. At the end of the day she places the labels on sheets of paper in the folders of the individual students. Her note on Tom is as follows:

> Tom   9/20
> Join-change unknown
> direct modeling
> didn't keep start and change separate

The note contains the student's name and the date, the problem type, the student's strategy, and a brief statement to remind her of the student's difficulty.

Ms. Grainger typically makes notes only when a student makes an unexpected response or seems to be showing growth. These notes help her revise her knowledge of those students. This form of assessment happens daily, but since children don't show daily growth, much of the assessment merely confirms what she already knows once her initial knowledge base is developed.

For about ten minutes Ms. Grainger watches and listens to all the students and quickly assesses whether her predictions about each student are being confirmed. As students complete the problem, they describe their solution quietly to Ms. Grainger. She observes that students who typically use direct modeling are using a direct-modeling strategy. She has previously seen some students using a "counting on" strategy. For this problem she expects those students to start counting at 20 and use fingers or counters to keep track as they count from 20 to 27. Ms. Grainger notes that the students who have previously used that strategy are using it on this problem.

Ms. Grainger doesn't organize instruction so that each lesson has a particular objective to be achieved, and she doesn't expect the children to learn something new every day. She does expect that each student will expand his or her ability to solve problems, make connections among mathematical ideas, reason mathematically, and communicate mathematical ideas effectively as a consequence of their daily interaction with appropriate mathematical tasks and of their communication with peers and the teacher about how they complete the tasks.

After the students have finished working on the problem, Ms. Grainger gives a few of them an opportunity to describe their strategy to the class. She used to let all the students describe their strategy but found that some strategies were described over and over and that the students didn't benefit from this repetition. Since she has already listened to many of the students describe their strategies to her, she can call on specific students who she knows will provide a variety of strategies. All the students want to share their strategies, and Ms. Grainger is careful to call on all of them regularly.

Tom describes his strategy. He tells how he used 19 counters to represent the 19 Kit Kat bars Ms. Grainger already has and how he then continued counting to 27, putting the additional counters in a separate pile so he could recount them. Most students have solved the problem this way.

Ms. Grainger doesn't want many repetitions of this strategy, so she purposely calls on students who used a different strategy. Students hearing another student describe a strategy different from the one they used may decide that they like the new strategy and begin to use it. This exchange of strategies is a common way for students to begin to use a more advanced strategy.

Ms. Grainger calls on Ellen to describe her "counting on" strategy, and then asks if anyone else used this strategy. Three other students, including Norman, raised their hands. Lisa reported her derived-facts strategy.

Anna did not solve the problem. She counted out 19 chips and then continued counting to 27, but she failed to keep the additional counters separate from the original 19 and thus couldn't solve the problem. Ted, Sara, and Sonia may be on the verge of moving from the use of direct-modeling strategies to the use of counting strategies. They seem to understand the counting-on strategy when other students describe their use of it. Ms. Grainger intends to create situations for them that will optimize this leap: she will have them solve join problems in which the first number is large and the second number is small.

This task is completed. Ms. Grainger has collected data from each student and has interpreted the data to extend her knowledge about each student's thinking. She has been making instructional decisions for individual students throughout the lesson, during each encounter with a student. What decisions might she make about the next task? Because she emphasizes problem solving and not practice, she has no reason to give the children another problem that is just like the one they just did. What are some of her options?

Options

1. Ms. Grainger could pose a similar problem that involves separating rather than joining.

2. She could pose a similar problem, but choose numbers that are closer together to increase the liklihood that more children will count on from the given number.

3. She could pose a similar problem with larger numbers but with a gap between the numbers that is either less than 10 (as in this problem) or greater than 10. This choice would promote the development of place-value concepts.

4. She could pose a join problem in which the initial number is unknown. An example: "Ellen needs 7 stickers to fill a page in her sticker album. She puts 12 stickers on each page. How many stickers are now on the page?" This problem is difficult for students who use direct modeling, as most of her students do, because it is not clear how to get started.

5. She could switch to a different type of activity, perhaps one involving graphing and comparison, measuring and graphing the outside temperature or the size of a sprouting plant, or some other activity not directly related to the activity just completed.

6. She could pose a problem that has a multiplicative structure, one that we would call a multiplication or division problem. Children who use direct modeling can frequently solve multiplication problems: "I have three packages of gum. Each package contains 6 sticks of gum. How many sticks of gum do I have?" They may also be able to solve measurement division problems. Ms. Grainger may want to use this opportunity to explore the children's understanding of some other problem types.

7. She could pose a comparable problem in a different context.

Other options are available as well. Ms. Grainger decides to pose a problem with numbers selected to facilitate "counting up to." Since there are twenty-three children in the class, she poses the following problem:

I want us to do a measuring activity with rulers. I have 21 rulers. How many more rulers do I need so that each of you can have a ruler to use?

SUMMARY

Children solve problems in well-identified ways. Teachers can learn to recognize how children solve problems, and they can learn to use that knowledge to make good instructional decisions.

The teacher's priority should be to attempt to understand how the students are thinking rather than to get the students to understand how the teacher is thinking. By constantly thinking about assessment during instruction, teachers can gain knowledge about the thinking of their students and create opportunities to extend each student's understanding. By thinking of instruction and assessment as simultaneous acts, teachers optimize both the quantity and the quality of their assessment and their instruction and thereby optimize the learning of the students.

REFERENCES

Bebout, Harriett, and Thomas P. Carpenter, "Assessing and Building Thinking Strategies: Necessary Bases for Instruction." In *New Directions for Elementary School Mathematics*, 1989 Yearbook of the National Council of Teachers of Mathematics, edited by Paul R. Trafton, pp. 59–69. Reston, Va.: The Council, 1989.

Carpenter, Thomas P., Elizabeth Fennema, P. L. Peterson, C. P. Chiang, and Megan Loef. "Using Knowledge of Children's Mathematics Thinking in Classroom Teaching: An Experimental Study." *American Educational Research Journal* 26 (1989): 499–531, 1989.

National Council of Teachers of Mathematics. *Curriculum and Evaluation Standards for School Mathematics*. Reston, Va.: The Council, 1989.

_____. *Professional Standards for Teaching Mathematics*. Reston, Va.: The Council, 1991.

Sammons, Kay B., Beth Kobett, Joan Heiss, and Francis Fennell. "Linking Instruction and Assessment in the Mathematics Classroom." *Arithmetic Teacher* 39 (February 1992): 11–16.

Webb, Norman, and Diane Briars. "Assessment in Mathematics Classrooms, K–8." In *Teaching and Learning Mathematics in the 1990s*, 1990 Yearbook of the National Council of Teachers of Mathematics, edited by Thomas J. Cooney, pp. 108–17. Reston, Va.: The Council, 1990.

Yackel, Erna, Paul Cobb, Terry Wood, Grayson Wheatley, and Graceann Merkel. "Social Interaction in Constructing Knowledge." In *Teaching and Learning Mathematics in the 1990s*, 1990 Yearbook of the National Council of Teachers of Mathematics, edited by Thomas J. Cooney, pp. 12–21. Reston, Va.: The Council, 1990.

4

Assessing a Wider Range of Students' Abilities

Malcolm Swan

I collect my students' work perhaps once each week, mark the answers for correctness, and then write a numerical or letter grade accompanied perhaps by a written comment at the bottom. When the work is handed back, students compare their grades, appearing more interested in who got the highest marks than in exactly where and why mistakes were made. I copy all their grades into my grade book (there usually isn't enough space for comments). At the end of each term and again at the end of the year, these grades are aggregated, averaged, and then presented in the form of a report to parents. This process takes hours (I have more than 200 students). Is it worth all the effort?

THE shortcomings of this common view of the classroom assessment process are all too evident. First, the purpose is confused. In the eyes of a teacher, for example, the purpose may be *formative,* recognizing the achievements of a student, pointing out errors and difficulties, and gathering information leading to the design of appropriate follow-up activities. Yet if feedback is provided in the form of aggregated scores, students will naturally perceive the purpose as being *summative,* that is, to rank the class in order of ability for an external audience, parents perhaps. Students' views (see fig. 4.1) can often be revealing.

The form of recording and reporting must be consistent with the purposes the assessment is designed to serve. Statistics become ends rather than means when the task of record keeping overrides the objective of helping students learn.

Second, the focus of attention is often too narrow. Assessment tasks that require students to imitate successfully isolated techniques through short, stereotyped questions set in artificial contexts cannot predict a student's ability to use mathematics autonomously on nonroutine problems taken from the real world. To do so requires a broader range of assessment practices, including the assessment of strategic skills. In addition, we may wish to assess conceptual

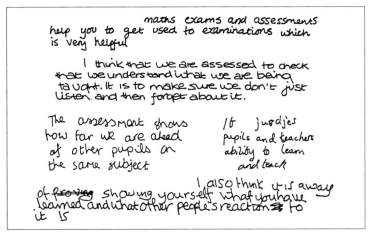

maths exams and assessments
help you to get used to examinations which
is very helpful

I think that we are assessed to check
that we understand what we are being
taught. It is to make sure we don't just
listen. and then forget about it.

The assessment shows   /t judges
how far we are ahead   pupils and teachers
of other pupils on   ability to learn
the same subject   and teach

I also think it is a way
of proving showing yourself what you have
learned and what other people's reaction to
it is

Fig. 4.1. Examples of students' comments. Reprinted from Ahmed et al. (1987, p.64).

understanding through tasks that require written or oral explanations. Perhaps we may even wish to assess a student's awareness and appreciation of the nature of mathematics and his or her attitudes toward the subject. A fuller description of these aspects is offered in figure 4.2. This agenda is now reflected in many curricular documents around the world.

This article illustrates the developments that have been made in the design of assessment tasks that reflect this broader range of objectives. Assessment for two different outcomes will be discussed: using mathematical and general strategies and achieving positive attitudes toward, and appreciation and awareness of, mathematics. In discussing these different outcomes, we shall explain and demonstrate forms of assessment that vary from summative tasks, which have been widely used in public examinations, to formative tasks, which have been used in only a few experimental classrooms. All the examples used have been tried and tested in classrooms in England.

ASSESSING THE ACQUISITION OF MATHEMATICAL AND GENERAL STRATEGIES

Mathematical strategies guide the choice of which skills are appropriate and enable pupils to use mathematics to tackle unfamiliar problems. Whereas a knowledge of facts and skills may be assessed through short, closed questions, the existence of strategic skills can be assessed only through more open tasks that require the students to make choices, reason, and explain. (Although this has hitherto been an unfamiliar area for assessment, the National Curriculum in England now specifies that all teachers must assess students' abilities to use and apply mathematical strategies at the ages of 7, 11, 14, and 16. This aspect is divided into three strands: making and monitoring choices, communicating mathematical ideas, and reasoning.)

Mathematics facts

Knowing notational conventions and correct terms for concepts

Examples:

- Remembering that (2, 3) indicates the point on a graph that is two units to the right of the y-axis and three units above the x-axis
- Knowing that the name given to the point (0, 0) is *origin*

Mathematics skills

Knowing how to carry through standard procedures

Examples:

- Doing long multiplication quickly and accurately
- Following rules for manipulating and solving equations

Mathematical strategies

Being able to approach nonroutine problems and use mathematical techniques in an automous way; knowing when particular methods may or may not be appropriate

Example:

Using such strategies as classifying, trying simple cases, symbolizing, finding helpful representations, being systematic, looking for patterns, making predictions and generalizations, and proving

Mathematics concepts

Understanding the meaning of mathematics concepts and how they relate to each other

Examples:

- Interpreting a graph and translating it into an alternative representation
- Explaining why $2 \div 3 = 2/3$

General strategies

Knowing helpful working habits and being able to use them autonomously

Examples:

Thinking globally, setting priorities, planning, using resources effectively, designing, organizing, and communicating

Appreciation and awareness

Being aware of—

- one's own current state of knowledge in the subject;
- types and purposes of various classroom activities (What is the main point of this lesson? What is the teacher trying to tell us?);
- the resources available for learning and how to use them (including the teacher, written of printed resources, or other students);
- ways of working and when they are appropriate (for example, individual styles of working for fluency practice on skills, group work, and discussion for developing understanding)

Personal attitudes and qualities

Developing a positive orientation toward the subject; becoming more confident, creative, outgoing, adaptable, cooperative, committed, and able to work on a team

Fig. 4.2. Aspects of learning that can be assessed

Short Tasks

Some aspects of mathematical strategy lend themselves quite well to assessment by short tasks, and such tasks have been used extensively in research studies and occasionally in public examinations. Three examples relating to the assessment of generalizing, symbolizing, and proving are given in figure 4.3.

The first of these, Skeleton Tower, has been used as a public-examination task. Students are able to use a number of different approaches. Some break the tower into four legs and a center, thus summing

$$4 \times (1 + 2 + 3 + 4 + \cdots + (n-1)) + n.$$

Others take horizontal slices:

$$1 + 5 + 9 + \cdots (n \text{ terms altogether})$$

A few others may even imagine breaking off two of the legs, turning them upside down and placing them on top of the others, thus forming a wall n units high and $(2n - 1)$ units long. The grading scheme gives credit under four headings: understanding the problem, organizing a systematic attack on it, explaining the results obtained, and generalizing in words or algebraically.

First to 100 is a simple number game that beautifully illustrates the power of inductive reasoning. After playing two games, the students in the extract below begin to reflect on winning strategies.

Theresa: When you get into the 80s, you can find a way to win. It's got something to do with 85 to 90.

(They play another game, and Theresa is on 88.)

Theresa: Now if I choose 1, I can win—I think.

(They play another game.)

Joanne: Look, if you get to 89, you must win, 'cause you couldn't win even if you chose 10.

In subsequent games, the students decided to stop at 89, since "the first to 89 wins." Then came the breakthrough. The argument was repeated—the first to 78 then wins. One student's explanation of the pattern they discovered appears in figure 4.4.

This example illustrates the capacity of these students to reason and communicate clearly. It can easily be developed into a more extended piece of work by allowing the students to change the constraints in some way, for example, by specifying that the first to 100 loses or that only numbers from 5 to 10 can be chosen.

In Roofs, the student must apprehend an unfamiliar notation, work with it, find conditions under which roofs may be drawn, and give explanations. This

Skeleton Tower

1. How many cubes are needed to build this tower?

2. How many cubes are needed to build a tower like this, but 12 cubes high?

3. Explain how you worked out your answer to part 2.

4. How would you calculate the number of cubes needed for a tower n cubes high?

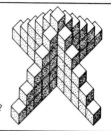

First to 100

This is a game for two players.
Players take turns to choose any whole number from 1 to 10.
They keep a running total of all the chosen numbers.
The first player to make this total reach exactly 100 wins.

So, in the sample game on the right, player 1 wins.

Play the game a few times with your neighbour.
Try to find a winning strategy.

Player 1's choice	Player 2's choice	Running Total
10		10
	5	15
8		23
	8	31
2		33
	9	42
9		51
	9	60
8		68
	9	77
9		86
	10	96
4		100

Roofs

Roofs can be drawn in different shapes and
sizes, using the dots provided.
The first one is 2 3 5 3.
The second is 3 1 4 1.

(The first number tells you how many units to
draw in direction 1, the second in direction 2,
the third in direction 3, the fourth in direction
4.)

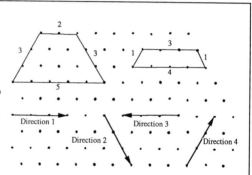

1. Draw a 2 2 4 2 and a 4 1 5 1.

2. Try to draw a 3 2 5 1 and a 1 4 3 4.
Explain what happens.

3. Find some rules which enable you to tell, *without drawing*, whether or not you can draw a roof from four given numbers.
Explain why these rules work.

4. If we call the four numbers a, b, c and d respectively, can you express your rules more simply?

Fig. 4.3. Three tasks designed to assess the acquisition of mathematical strategies. (Roofs is from Bell, Shiu, and Horton (1981); Skeleton Tower and First to 100 are from Swan (1984).

There is a winning pattern in this game.
If you start with a one then your
opponent will add a number between one
and ten. You then make the total up to 12.
Then they add a number the total of which
you make up to 23. Here is the pattern
of winning numbers

1 12 23 34 45 56 67 78 89 100
(each time the difference is eleven)

If you carry on this pattern your opponent
cannot win as they cannot add eleven
to make the total up to one of the winning
pattern numbers.

Fig. 4.4

task can be used to assess the ability of students to reason, communicate, and make generalizations using algebra. The conditions sought are $a + b = c$ and $d = b$. Students can argue the truth of these either empirically by looking at a number of usable codes or structurally or deductively by looking at features of the diagram. These approaches are equally acceptable.

Extended Tasks

Although the foregoing tasks are useful in assessing mathematical strategies, they are essentially straightforward in that they are short, well defined, and limited in scope. During the last few years, much attention has been paid to the use of extended, more open tasks that can typically occupy from three to fifteen hours of class time. These tasks are now in extensive use throughout England. In this environment, students are more able to generate their own questions for investigation and explore their own lines of reasoning.

Two examples of such tasks are given in figure 4.5. The first, Consecutive Sums, is a pure number investigation; the second, Design a Shape Sorter, is an extended real problem-solving activity. Here, *real* is taken to imply that the students actually put their solution into practice. Notice that the students have the opportunity to pursue their own questions and design their own products, thus giving them a sense of ownership of the task, which is the key to their being able to sustain effort over many hours.

One fifteen-year-old girl, for example, decided to design and make a shape sorter in the form of a house. She began by sketching several designs and evolved criteria that her design should satisfy: each shape should pass through only its own slot, and the means of retrieving shapes after they have been

CONSECUTIVE SUMS

The number 15 can be written as the sum of consecutive whole numbers in three different ways:

$$15=7+8$$
$$15=1+2+3+4+5$$
$$15=4+5+6$$

The number 9 can be written as the sum of consecutive whole numbers in two ways:

$$9=2+3+4$$
$$9=4+5$$

Look at numbers *other than* 9 and 15 and find out all you can about writing them as sums of consecutive whole numbers.

Some questions you may decide to explore . . .

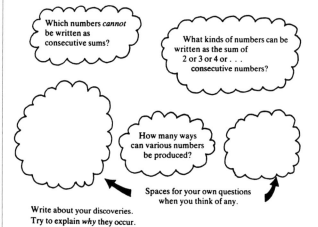

Which numbers *cannot* be written as consecutive sums?

What kinds of numbers can be written as the sum of 2 or 3 or 4 or . . . consecutive numbers?

How many ways can various numbers be produced?

Spaces for your own questions when you think of any.

Write about your discoveries. Try to explain *why* they occur.

Design a shape sorter

Young children are often given "shape sorters" to help them develop eye and hand coordination.

What do you think makes a good shape sorter? Write down your list of desirable features.

Design your own shapes and shape sorter and make them all out of cardboard. Write about how you designed it and why you designed it as you did.

Give your shape sorter to young children and watch them play with it. Which shapes do they find easy to put in? Why? Which shapes are harder? Why?

Fig. 4.5. Two "starters" for extended tasks. Consecutive Sums is taken from Swan (1984); Design a Shape Sorter is taken from Swan, Binns, and Gillespie (1988).

inserted should be simple. After designing nets for the house and the solids that would be inserted, she then made a creditable model. (See fig. 4.6.)

Fig. 4.6. A shape sorter designed by a fifteen-year-old girl

Of course, considerably more support and guidance may need to be given than indicated here. The art of designing effective support involves obtaining a balance between giving sufficient structure so that the students are not bewildered and frustrated by the demands of the task and offering too much guidance so that the strategic load is effectively removed. We cannot assess strategic skills if we make all the decisions for the students!

Recording Evidence of Strategic Skills

Grading schemes for tasks that assess strategic skills are necessarily of the profile type. One such scheme used by the largest examination board[1] in England, for example, awards marks under each of the headings below (paraphrased from the 1991 coursework criteria of the Northern Examining Association (NEA), Manchester, England [NEA 1990]):

1. Understand and respond to a task

   This includes assessing the ability of the student to adopt a suitable strategy, break down a multistage task into identifiable stages, identify realistic goals, and choose appropriate equipment.

2. Reason and make deductions

---

1. In England, five examination boards, associated with groups of universities, work with teachers under the direction of the Government Department of Education to devise and run examinations for students at the ages of sixteen and eighteen. Schools are free to choose from the syllabi they provide.

This includes assessing the ability of the student to identify patterns and generalize, draw and justify conclusions, and consider variations and extensions to a task.

3. Work on a task

This includes assessing the ability of the student to collect and process appropriate data, carry out calculations to appropriate degrees of accuracy, overcome difficulties, organize work, and verify work and results.)

4. Use equipment

This includes assessing the ability of the student to use software packages, calculators, measuring and geometrical apparatus, scissors, string, straws, glue, and so on, as appropriate.

5. Estimate and make mental calculations

This includes assessing the ability of the student to make realistic and sensible approximations and estimations, when calculating with physical quantities.

6. Communicate

This includes assessing the ability of the student to present a clear and orderly written report of his or her work, communicate orally the processes and results, and discuss mathematical ideas.

Teachers are required by the examination board to award marks on an eight-point scale. Naturally, not every piece of work is expected to be assessable under every heading, but over a period of time a student's portfolio is expected to provide evidence of attainment under each criterion. A teacher's own professional judgment must be used to adjust the marks according to how much help the student needed. When these assessments are to be made summatively, for public use, teachers also need to meet together to ensure that their standards are comparable, which occurs at an examination board's "moderation" meetings.

Of course, tasks that span several weeks often involve collaborative group work, and teachers are often concerned about how to assess an individual's attainment if she or he has been working cooperatively. Possible solutions include the use of oral interviews and the occasional introduction of related tasks (sometimes called "controlled elements") that must be completed individually.

Many schools have found it helpful to make students explicitly aware of such grading schemes. Indeed, it is worthwhile to encourage students to evaluate examples of extended pieces of work for themselves. This kind of reflective activity can develop a student's critical faculties—an important educational aim in itself. More examples of this kind of activity are described in the following section.

ASSESSING ATTITUDES, APPRECIATION, AND AWARENESS

Often, students engage in mathematical activity without really being aware of the purpose of the activity or what they have learned. Working *through* tasks becomes a substitute for working *on* ideas. Recently, we have been exploring ways of increasing the quantity and quality of reflective activity that takes place in the classroom with the intention that students will become more responsible for their own learning.

A teacher can use several means of encouraging students to reflect. Some involve the students' changing their role and thus their point of view. The following two examples relate to assessment:

Students' Assessing Themselves

This secondary school organizes its year-7 curriculum (for 11- and 12-year-old students) around six extended projects. Each project occupies about six weeks of classroom time and involves the students in working in groups on such holistic tasks as planning a classroom layout, in which the students assess existing layouts, measure the existing furniture, make scale models, and implement the preferred proposals.

At the start of each project, students are given a list of about ten performance criteria and occasionally during the project are encouraged to record on a five-point scale their perceptions of their own progress according to these criteria. This assessment is done in an informal way; it seems that sometimes the scale is interpreted as the "level of understanding" and other times as the "level of autonomy" that was evident.

The criteria and self-assessment sheet in figure 4.7 refer to a geometry project using Logo. The girl in this example has assessed herself as scoring 4 out of 5 on criterion 1, understanding and using language associated with angle. Her teacher, believing that her work rates a better score, has given her 5 out of 5. Where possible, the teacher explains why his or her assessment differs from the student's.

At the end of the project, students are also asked to rate their own ability to comprehend, plan, carry out, and communicate their work. Using the same scale, the teacher then writes her or his own complete evaluation alongside the student's. Considerable discussion and negotiation follows between pupil and teacher, with the intention that both parties become more aware of areas of progress and particular needs. It should be noted that students are also encouraged to write about what they think they have learned, how hard they judge they have worked, and how much they have enjoyed the topic.

Any work of which the student is particularly proud is attached to the self-assessment sheet and preserved in a filing cabinet. Thus a portfolio of work is collected, which the student is responsible for keeping up to date.

Criteria

1. I can understand and use language associated with angle.
2. I can construct simple two-dimensional shapes.
3. I know the angle properties of quadrilaterals.
4. I can use Logo to transform 2-D shapes.
5. I can recognise
6. I can use coordin
7. I can recognise r

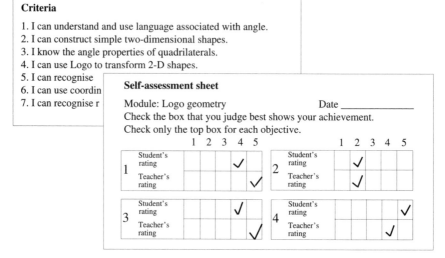

Self-assessment sheet

Module: Logo geometry   Date _____
Check the box that you judge best shows your achievement.
Check only the top box for each objective.

Fig. 4.7

This whole process of the student's involvement leads to a better understanding between the teacher and the student and a greater awareness of what is being learned.

Students' Assessing Each Other

In another school, a teacher asks a year-9 class (thirteen- and fourteen-year-old students) to devise their own test at the end of a unit on graphs. Students begin by listing the most important ideas in the topic together with the different types of questions that can be asked. They are encouraged to devise questions of varying difficulty that are interesting and that they know they are capable of answering. Students also supply their own answers and grading schemes. Items from individuals are collected and used to form a composite test for the whole class.

This type of activity forces students to review the topic and itemize the most important facts, concepts, and skills. Students also reflect on the design of questions and on the limits of their own capabilities. Often the resulting test is much harder than the teacher expects, and the students perform better than anticipated.

Assessing Students' Attitudes and Perceptions

The extracts in figures 4.8 and 4.9 illustrate further tasks that are currently being used in research to assess students' attitudes toward mathematics and their perceptions of the purposes of several types of activities. In their current

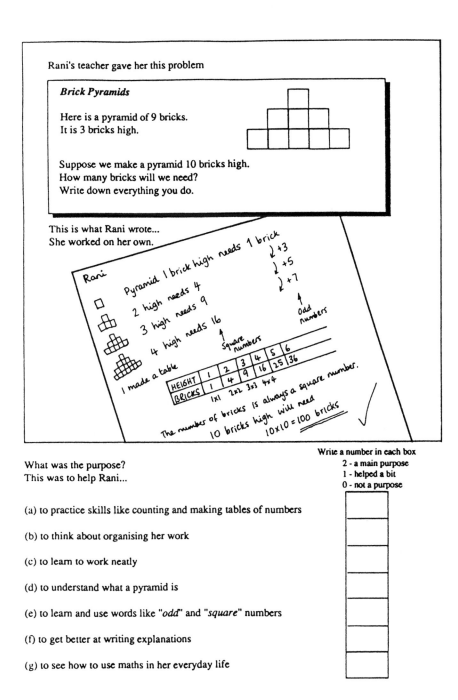

Rani's teacher gave her this problem

Brick Pyramids

Here is a pyramid of 9 bricks.
It is 3 bricks high.

Suppose we make a pyramid 10 bricks high.
How many bricks will we need?
Write down everything you do.

This is what Rani wrote...
She worked on her own.

Rani

Pyramid 1 brick high needs 1 brick
2 high needs 4
3 high needs 9
4 high needs 16

I made a table

HEIGHT	1	2	3	4	5	6
BRICKS	1	4	9	16	25	36

1×1 2×2 3×3 4×4

square numbers

odd numbers
↳ +3
↳ +5
↳ +7

The number of bricks is always a square number.
10 bricks high will need
10×10 = 100 bricks

What was the purpose?
This was to help Rani...

Write a number in each box
2 - a main purpose
1 - helped a bit
0 - not a purpose

(a) to practice skills like counting and making tables of numbers

(b) to think about organising her work

(c) to learn to work neatly

(d) to understand what a pyramid is

(e) to learn and use words like "*odd*" and "*square*" numbers

(f) to get better at writing explanations

(g) to see how to use maths in her everyday life

Fig. 4.8. A sample task attempting to assess a student's perceptions of the purpose of a mathematical activity. Reprinted from a current research project at the Shell Centre for Mathematical Education, Nottingham University, England

Learning maths is like...

Tick one box for each statement

strongly agree
agree
not sure
disagree
strongly disagree

(a) Learning maths is like learning a new cooking recipe.
The teacher or book gives you step-by-step instructions.
You just do what they say.

(a) ☐☐☐☐☐

(b) Maths is like a jungle.
The ideas are all jumbled up.

(b) ☐☐☐☐☐

(c) Doing a maths problem is like crossing a river on
stepping stones.
There is only one way to go.

(c) ☐☐☐☐☐

(d) Learning maths is like building a wall.
You have to lay the bricks in order.
You have to learn maths ideas in a certain order.

(d) ☐☐☐☐☐

(e) You don't need to understand how maths works, you
just need to practice doing it.

(e) ☐☐☐☐☐

(f) Learning maths is like exploring an unknown country.
You make lots of choices, where to go, what to do.

(f) ☐☐☐☐☐

(g) Maths is like a jigsaw.
The ideas fit neatly and beautifully together.

(g) ☐☐☐☐☐

(h) Doing a maths problem is like finding your way through
a maze. There are lots of possible pathways to go down.

(h) ☐☐☐☐☐

(i) Learning maths is like drawing a picture. It doesn't
matter which bit you do first. It will all fit together in
the end.

(i) ☐☐☐☐☐

(j) ☐☐☐☐☐

(j) You need to understand each idea in maths before you
use it.

Fig. 4.9. A sample task attempting to assess a student's perceptions of mathematics in general.
Reprinted from a current research project at the Shell Centre for Mathematical Education,
Nottingham University, England

form they are rather closed; perhaps more open responses would be more informative, but nevertheless they have yielded interesting reponses. Some teachers are beginning to acquire the habit of asking their students to note at the end of a series of lessons their views on the purpose of the lessons, what they believe they have learned, how hard they think they have worked, and how much they have enjoyed it.

CONCLUDING REMARKS

Regretfully, externally designed, summative-assessment practices often largely determine what is to be taught and the style of the internal formative-assessment procedures adopted by teachers. External examinations always seem to place a premium on the fluent performance of technical routines in stereotyped contexts, perhaps partly because these aspects are the easiest to mark. The thesis of this article is that the range of assessment tasks should be broadened to reflect the wider range of achievements we value. This challenging problem in design is complex, but we must face it if our assessments are to have any validity and formative value.

REFERENCES

Ahmed, Afzal. *Better Mathematics, a Curriculum Development Study.* London: Her Majesty's Stationery Office, 1987.

Bell, Alan, Christine Shiu, and Bruce Horton. *Evaluating Attainment in Process Aspects of Mathematics,* Nottingham, England: Shell Centre for Mathematical Education, University of Nottingham, 1981.

Northern Examining Association. *Mathematics through Problem Solving.* A GCSE Syllabus, Manchester, England: NEA, 1990.

Swan, Malcolm, ed. *Problems with Patterns and Numbers, an Examination Module for Secondary Schools.* Manchester, England: Shell Centre for Mathematical Education, University of Nottingham, and Joint Matriculation Board, 1984.

Swan, Malcolm, Barbara Binns, and John Gillespie. *Numeracy through Problem Solving.* Five modules: "Design a Board Game," "Produce a Quiz Show," "Plan a Trip," "Be a Paper Engineer," "Be a Shrewd Chooser." Harlow, England: Shell Centre for Mathematical Education, University of Nottingham, Joint Matriculation Board, and Longman Group, 1988.

5

Classroom Assessment in Japanese Mathematics Education

Eizo Nagasaki
Jerry P. Becker

FOR several decades, assessment has been discussed from various points of view in Japanese mathematics education. For example, the following questions have been discussed:

How are the grading (rating) and assessment of student performance related?

How are interest in, and attitudes toward, mathematics assessed?

Should we use criterion-referenced or norm-referenced assessment for grading purposes?

How can we cope with the effects of the entrance examinations as a part of external assessment?

Sometimes the media (e.g., newspapers and magazines) have functioned as a forum in the debate about the merit(s) or demerit(s) of assessment, especially with respect to the effects of the entrance examinations. Indeed, ordinary people have also been involved in the controversy surrounding the entrance examinations. Also, more recently, criterion-referenced assessment has become an issue along with the question of how to assess (*a*) mathematical thinking and (*b*) interest in, and attitudes toward, mathematics. Japanese mathematics educators have been struggling with the problem of how to use both entrance examinations and criterion-referenced assessment to measure mathematical thinking and interest in, and attitudes toward, mathematics.

This article was a fully and equally shared effort between the two authors. It stems from the second author's interest in Japanese mathematics education, dating back many years, the U.S.–Japan Seminar on Mathematical Problem Solving (Becker and Miwa 1987), and his professional interaction with the first author and many of his colleagues. The article follows on the heels of the completion of the translation of the book on the "open-end approach," which was edited by Shigeru Shimada. The authors express their appreciation to Professor Shimada for his very helpful comments on a draft of the paper and also to Norman Webb for his very useful suggestions and patience in completing the task.

THE MATHEMATICS CLASSROOM
IN JAPANESE SCHOOLS

Classroom Teaching

Whole class instruction is the approach used by mathematics teachers in Japanese elementary and secondary school classrooms. All classrooms are equipped with a large chalkboard on the front wall, many with an overhead projector and screen. Some teachers also use small 2' × 2' or 3' × 3' chalkboards (which are hung from the top of the large chalkboard) or poster boards attached to chalkboards. Teachers use these objects for the presentation of problems and solutions, or at times the teacher has students write their problem solutions and approaches on them for display to the whole class.

Class size in Japan is much larger than in the United States. Typically there are about thirty to forty-five students in a class at the elementary and secondary school levels. They sit in a boy-girl configuration at desks with benches (sometimes in rows of single desks). Students, especially in elementary schools, are quite disciplined and attentive during class and also somewhat formal compared with their cohorts in the United States. For example, a lesson begins and ends with students rising and bowing to the teacher, and the teacher reciprocates. A similar situation prevails in high schools, though at present some high schools are experiencing discipline problems with students. Since there are ten minutes between classes, students have an opportunity to relax and "unwind" after intensive concentration during class; at times, the teacher may extend the class period in order to complete and polish the lesson.

Generally the teacher develops the lesson around *one single objective* (e.g., a topic or behavioral objective), and class activities are *focused on it*. The main role of the teacher is that of a guide, not a "dispenser of knowledge." The activities and sequence of events in a lesson are commonly organized to draw out the *variety of students' thinking*. The teacher's "wait time" is crucial in this respect. The different ways students think about the mathematical topic or problem in a lesson are respected to a very significant degree by the teacher; in fact, the dynamics of a lesson center on this, and *teachers rely on students as an "information source"* during the lesson. The *discussion* of students' ideas is also a prominent characteristic. Similarly, students are expected to give verbal explanations, sometimes lengthy ones, of their ideas (cf. Stigler 1988). Toward the end of the lesson, the teacher "pulls together" students' ways of thinking, discusses their mathematical quality, and then summarizes, elaborates, or "polishes up" the lesson. Discussion (whole class or small group) among students or between the teacher and students is *extensive* and is a major factor in achieving the lesson's objective (cf. Becker et al. 1989, Becker et al. 1990, Stigler 1988, Stigler and Stevenson 1991, and Miwa 1992).

Lessons are also *intensive*. The lesson moves toward the objective with minimal external interruptions (e.g., public address announcements or students entering or leaving the room). There is a certain discipline about this, in which students' own responsibility in learning is reflected. Boy-girl interaction is common, and teachers have high expectations of both sexes.

What goes on in the *teachers' rooms* is another important characteristic of Japanese education. Unlike in the United States, teachers in Japan share a large room, each with a desk, chair, and file. This arrangement is no trivial detail, for it gives teachers an opportunity to interact: they plan lessons, discuss written records of teaching, discuss and plan evaluation, and, in general, discuss professional matters relating to their students, teaching, and the mathematics curriculum.

The Role of Lesson Plans

The philosophy and reality of lesson plans and lesson records of teaching in Japan are considerably different from those in the United States or some European countries. Regarding lesson plans, many American teachers may think it impossible to anticipate students' responses in detail; therefore, their lesson plans may be somewhat rough or not detailed. Many Japanese teachers, however, think that it is crucially important to develop and polish lesson plans in a collaborative manner, including listing students' anticipated responses to the problems posed in a lesson (see fig. 5.1). In this way teachers get a better understanding of a lesson themselves, and they are better prepared to anticipate and deal with students' responses and viewpoints in the actual teaching. A typical Japanese classroom lesson may be compared to a drama, with the lesson plan the script (see Yoshida 1992).

A PERSPECTIVE ON ASSESSMENT IN JAPAN

Approach to Assessment

Within the framework of whole-class instruction, many Japanese teachers respect classroom teaching that is directed toward using different ways of students' thinking in order to raise the level of the students' understanding in the class as a whole. Therefore, the focus of assessment is on "how each student thinks according to her or his natural way of thinking or ability." These ways of thinking mathematically are regarded as concrete information about students' progress in learning. Both correct and incorrect ways of students' thinking are naturally included. It is believed that using all the ways in a lesson helps to enhance students' learning.

In this approach, teachers' observations of students during the lesson are an important source of information for assessment. According to their observations, teachers adjust their teaching, for example—

• to see how well students understand their task;

Objective: To help students solve the problem by drawing out students' natural ways of thinking, comparing them, and finding a rule

Teaching

1. The problem:

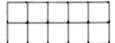 If matchsticks touch only at their endpoints as in the figure, how many matchsticks do we need to make 10 squares? Find the answer in as many ways as possible.

2. Introducing the problem:

Teacher: How many matchsticks do we need when we have two squares?
Student: Seven matchsticks
Teacher: What about when we have four squares?
Student: Twelve matchsticks
Teacher: Now find the answer to the problem in as many ways as possible. Think out a variety of ways for determining the number of matchsticks. Show your work on your worksheet.

3. Students' anticipated responses:

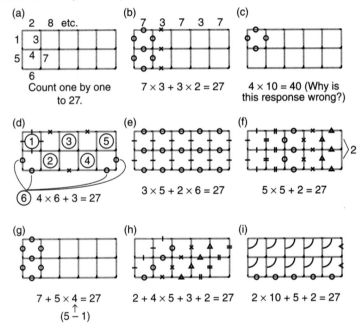

(a) Count one by one to 27.

(b) $7 \times 3 + 3 \times 2 = 27$

(c) $4 \times 10 = 40$ (Why is this response wrong?)

(d) $4 \times 6 + 3 = 27$

(e) $3 \times 5 + 2 \times 6 = 27$

(f) $5 \times 5 + 2 = 27$

(g) $7 + 5 \times 4 = 27$
$(5 \overset{\uparrow}{-} 1)$

(h) $2 + 4 \times 5 + 3 + 2 = 27$

(i) $2 \times 10 + 5 + 2 = 27$

(j) Others

Note: "Two times three" is written 3×2 in Japan.

4. Discuss each way with students and classify them according to some shared feature. Compare the different ways according to their mathematical quality.

(a) Which way do you think is best? Why?
(b) What happens when the number of squares increases? Explain.
(c) Which is the easiest way when we have 20 squares?

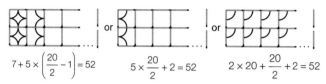

$$7 + 5 \times \left(\frac{20}{2} - 1 \right) = 52 \qquad 5 \times \frac{20}{2} + 2 = 52 \qquad 2 \times 20 + \frac{20}{2} + 2 = 52$$

5. Generalize: $7 + 5 \times (n \div 2 - 1) = $ # of matchsticks
                         $5 \times (n \div 2) + 2 = $ # of matchsticks
                         $2 \times n + (n \div 2) + 2 = $ # of matchsticks

6. Summing up

7. Homework: How many matchsticks do we need when we have the following
   shapes?
   (a) 7 squares
   (b) 15 squares

Fig. 5.1. A teacher's plan for a lesson on problem solving in grade 6 (edited from the Japanese)

- to select which response(s) will be presented to the whole class;
- to enhance the quality of discussion;
- to pay attention to students' individual needs.

Teachers make two types of observations: observations of students' work on the problem while walking around the room, and observations made during discussions with students. Included in the first type is observing whether students' responses are as anticipated. After the lesson, students' worksheets are collected and analyzed as another crucial source of information to evaluate both the lesson and individual students' performance.

Observation as an approach to assessment is accepted as important by many teachers at the elementary school level; the higher the grade level, however, the fewer the number of teachers who actually use this approach. One main reason is that they do not have enough time to make observations. Another reason, especially for high school teachers, is the importance of, and the time devoted to preparation for, entrance examinations (at grades 9 and 12). Therefore, more formal assessment, which depends on paper-and-pencil tests, is generally the main approach in classrooms at all school levels. This trend is especially strong in senior high schools.

At the elementary school level, teachers understand that classroom assessment needs to be integrated into classroom lessons. This may lead, in a natural way, to curriculum improvement based on classroom practice. In Japan, this concept is expressed by the slogan, We Should Learn from the Students.

Assessment Practice

In this section, assessment both during and after lessons is considered. Several lesson records made by teachers after they taught the lessons are used to show some aspects of assessment. Here we focus on lessons in which the teachers try to use students' different methods of solving a problem, which reflect their ways of thinking, to assess their students' learning.

Assessment during lessons

Classroom lessons mainly involve whole-class instruction, as mentioned earlier, and the chief means of assessment is teacher observation. Here we see from lesson records how assessment during lessons is implemented. It is important to observe how students think. In this stage, an assessment of classroom instruction is the main objective, but if observations of each student are accumulated, the results of the assessment will become information for the summative evaluation of each student.

1. Assessment of concept formation in the whole class. In whole-class instruction, students' different ways of thinking can be used to form a concept from different points of view. It is crucial, however, first to formulate a problem situation in which every student can have some success in finding some solution method(s). After students exhibit their ways of thinking, the teacher should classify them to form a concept.

Example—Number of matches: fifth grade (Hashimoto 1989)

The teacher, Mr. Tsubota, presented the following problem:

Squares are made using matches as shown in the figure below. When the number of squares is 5, how many matches are used?

Students presented their methods of finding the answer as follows:

• Drawing the figure and counting one by one (important for less able students)

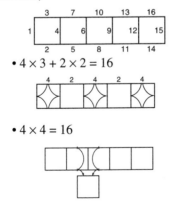

• $4 \times 3 + 2 \times 2 = 16$

• $4 \times 4 = 16$

• $2 \times 5 + 6 = 16$

• $3 \times 5 + 1 = 16$

• $4 + 3 \times 4 = 16$

• $1 + 3 \times 5 = 16$

• $4 \times 5 - 4 = 16$

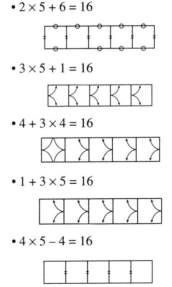

The teacher could see that students understood this problem very well, for they developed numerous methods for finding the answer. Since students looked at this problem from several viewpoints, the teacher could now proceed to the next stage, namely, asking students to make up, formulate, or pose new problems by themselves. So the next lesson began with a review of the last lesson, including reference to the eight ways students used to get the answer. Excerpts of the remainder of the record of the whole lesson follow ["___-*kun*" denotes boys, "___-*san*" denotes girls]:

T: In today's lesson, I won't pose a problem, but you will pose one by making up a problem similar to the one you just solved. I want you to present the problem you made yourself and discuss it with one another.

S: Is it okay if it's only a little bit similar?

T: Yes, it's okay.

S: Is it okay if we use triangles or pentagons instead of squares?

T: That's a good idea, but if you say your ideas out loud, others may end up using them. Let's begin.

S: Teacher, may I draw a figure?

T: Yes; if you draw a figure, it will make it easy for others to understand your problem.

[Teacher walks around, scanning students' work.]

. . .

T: Sonobe-kun, please come up and explain your problem.

S: I changed the first problem a little, and made this problem:

Squares are made using iron sticks. If the number of squares is 30, how many iron sticks are used?

T: What is the length of all the sticks? Are they the same length?

S: Constant length sticks.

. . .

T: Did anyone make a problem similar to this? Shoji-kun?

S: My number is different. Seventy sticks.

T: How many people changed the number of squares?

[Ten children raised their hands.]

. . .

T: Tani-kun, please explain your problem. Listen to his idea, everyone.

S: Squares are made by matches in the first problem. I made the problem different by changing squares to equilateral triangles like this:

Equilateral triangles are made by using matches as shown in the figure. When the number of equilateral triangles is 15, how many matches are used?

T: Did you only change squares to equilateral triangles?

S: I also changed the number.

T: Raise your hand if you changed squares to triangles.

[Many children raised their hands.]

T: Oh, so many. Well, how many people changed squares to geometrical figures other than triangles?

T: What figures did you make, Endo-kun? Come up and put yours on the blackboard.

S: Well, I changed squares in the first problem to regular hexagons, and I changed the number from 5 to 1011:

Matches are arranged as shown in the figure. When the number of regular hexagons is 1011, how many matches are used? (Use matches of all the same length.)

T: Can you solve it? I think you can.

S: It's solvable if I compute it. Maybe I can.

. . .

T: There are two types so far—those formed by changing squares to equilateral triangles and those formed by changing squares to regular hexagons.

T: Did anyone pose the problem by changing to other figures besides the hexagon? What did Suzuki-san do?

S: I want to make four pentagons with five beads per side. How many beads are used?

T: Please draw your figure. By hand is okay.

. . .

T: Triangle, hexagon, pentagon. Did anyone make other figures? Yes, Kosaka-san.

S: Rectangular solid.

T: Rectangular solid? Did you draw it? That's interesting.

T: Tsunashima-kun?

S: Rectangle.

T: Draw your figure.

T: Suzuki-kun? Write your idea above Suzuki-san's response.

[Suzuki-kun drew the two left figures first, and then moved to Suzuki-san's response in the right figure.]

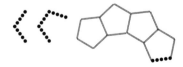

T: You just fill in one side of beads with yellow chalk, we can understand. Please explain.

S: I want to make four pentagons with 5 beads per side. How many beads are used?

. . .

T: She made such a problem…the figure is a pentagon. Thanks. Any questions? Ariga-kun, okay?

S: Yes. A vertical figure of regular pentagons is made using matches. When the number of regular pentagons is 726, how many matches are used?

T: Regular pentagons are connected like this. It's different from the first problem, because in his problem, the figure is zigzag while the first one isn't. Thanks.

. . .

T: Then Tsunashima-kun. What is your problem?

S:

Rectangles and squares are made by using matches as in the figure. When the number of rectangles and squares altogether is 1111, how many matches are used? [One side of the rectangle is two times that of the square.]

T: I don't understand the meaning of the figure.

S: I know. Rectangle, square, rectangle and square. [Tsunashima-kun calls on Ariga-kun, who has raised his hand.] If one side is doubled, is each side of the rectangle doubled? Are both width and height doubled?

S: Only the width is doubled.

T: Well, one more person, Suzuki-kun, came up with a different way of posing the problem.

S: Yes. I almost completely changed the problem. And this is the problem. Parallelograms are made by using pencils of the same length as shown in the figure below. When the number of pencils is 37, how many parallelograms can be made and how many pencils will remain?

T: How did you change it? You said you almost completely changed it.

S: Yes. I changed matches to pencils and squares to parallelograms. And instead of asking how many matches make squares, I asked how many parallelograms can be made with 37 pencils and how many pencils will remain.

T: It seems a little difficult, but any questions?

S: What if there is no remainder? It's okay.

...

[There ensues a discussion of what a square is, so it isn't a square (no 90° angles). It's a rhombus, since the pencils are the same length, but it is also a parallelogram. There is also discussion of division in finding the answer. Then there was a fairly lengthy discussion (teacher-students and students-students) about how the posed problems were the same or different, and in what ways (e.g., ◁▷ and △▽△).]

...

T: Well, time is up, but the first method by Sonobe-kun is increasing the number of squares, and the methods by Tani-kun, Endo-kun, Kaneko-san, Suzuki-san, and Tsunashima-kun involve changing the figures. Of course, the number is also changed.

[The teacher pointed to two problems like Suzuki-kun's second problem; i.e., the converse problem.]

T: These two interesting problems are different from the others—they give the number of matches and ask how many figures can be made.

Each problem belongs to one of three types. In the first type, the number is changed, that is, the number of squares. In the second type, the figure is changed. The third type is the converse problem. What type of problem would you want to solve if you were to solve one of these problems?

S: Endo-kun's problem.

T: And you?

S: Endo-kun's problem.

The answer for the original problem is unique, but there is a rich variety of methods or ways of thinking that students use to find the answer. The teacher can see some cognitive development by observing individual student findings and, through discussion, how the findings of students are similar or different and classifying the results into categories.

2. Assessment of individual concept formation. Especially in whole-class instruction, attention should be paid to individual students; this is particularly true for teaching basic ideas in mathematics. Teachers must grasp students' thinking and deal with each one individually. In doing so, it is extremely important that teachers try to *anticipate* all possible ways of students' thinking and consider the methods that may be used to deal with them beforehand.

Example—Number up to 100: first grade (Matsubara 1987)

T: There are many marbles in this box. When I take a handful of marbles, how many do I have?

S1: 50 marbles.

S2: 100 marbles.

S3: 84 marbles.

T: How many marbles can you take?

S4: 40 marbles.

S5: 100 marbles.

S6: My teacher, let's take a handful!

T: Okay, please come take a turn.

[Students take 30 to 50 marbles.]

T: Who took the most marbles? Please count them. How many marbles did each of you take? Please arrange them in such a way that I can understand.

[Even before I asked them to count, they began. Walking around the classroom, I observed how the students were counting. I asked for their methods, one by one.]

T: [To a student who counts one by one] How many marbles are there?

S7: 41 marbles.

T: I am not sure whether the number of marbles is 41 or not, unless I count one more time.

T: [To a student who counts by tens, with 10 marbles in a line] How many marbles did you take?

S8: 35 marbles?

T: How did you do it?

S8: Here, 1, 2, 3, ..., 10, there are 10 marbles. Since there are three 10s, that's 30 marbles, and 5 marbles remain. In total, 35 marbles.

T: Well, you arranged the marbles in a good way.

T: [To a student who makes a pile of marbles] What is that pile?

S9: I separated 10 marbles.

T: Please show me whether the number of marbles is really 10 or not.

S9: 1, 2, 3, ..., 9. Oh, there are only 9 marbles here!

T: Please make a group in such a way that you can see the marbles clearly.

...

Finally, the teacher found five ways of counting by the students:

• Counting one by one
• Counting by piles of ten
• Counting by ten in a triangle
• Counting by five in a line, ten in two lines
• Counting by ten in a line

The teacher assesses each individual student and acknowledges each approach used. Therefore, the teacher could determine that all the children were ready for the place-value system.

3. Assessment during discussion. Discussion between students and the teacher is usually undertaken after students individually tackle a problem. In the discussion, students can see firsthand that different opinions among them exist and can then recognize concepts more deeply. The teacher's questions and observations together promote the discussion. Observation during students' work on a problem becomes an important source of information for questions.

Example—Linear equations: seventh grade (Handa 1992)

The teacher presented the following problem:

When do the long hand and the short hand overlap each other on a clock between 1 and 2 o'clock, and 2 and 3 o'clock?

The students thought freely about this situation. Then the teacher observed students' responses by walking around and classified them roughly as follows:

• Many students made an equation.

• Some students solved the problem without making an equation.

• A few students drew a figure.

The teacher started the discussion.

T: Please present how you solved the problem. [Intentionally, the teacher named a student who solved the problem without making an equation.]

S1: In elementary school, I had solved it by using a mathematical expression. The long hand proceeds 6° in a minute. Since the short hand proceeds 0.5° in a minute, the long hand overtakes the short hand $(6 - 0.5)° = 5.5°$ in a minute. Therefore, to overtake 30°, it takes $30 \div 5.5$, namely, 60/11 minutes. The first overlap is at 1 o'clock 5 5/11 minutes. I thought the same way for the next problem.

T: Are there any questions for this way of thinking? No? Then how can we solve the situation between 2 and 3 o'clock by this way? [Teacher calls on another student.]

S2: At 2 o'clock, there is a 60° difference at first, $60° \div 5.5°$, or 11 minutes.... Just a minute. [She started to reduce 600/55.]

S3: Since it is twice, it is 120 elevenths.

T: You said it was twice; how did you think of that?

S3: The first difference, 60°, is twice as much as 30°.

T: I see; if so, how can we calculate the time the long hand and the short hand overlap at past 3 o'clock? [Teacher calls on a different student.]

S4: Since it is three times, it is 180/11 minutes.

T: In the same way, for 4 o'clock, 5 o'clock, to 60/11, we do four times, five times....

T: Now, we will let students who solved the problem using other methods present theirs. [Teacher intentionally named a student who solved the problem using an equation.]

S5: I solved by making an equation. After all, it's the same.

T: You are right. If it were different, it would trouble us (laughing). What equation did you make?

...

The teacher used two types of observation in this situation: observation during the *tackling* of the problems and observation during the *discussion*. The teacher's purpose was reflected in selecting which students to respond. The teacher wanted the students, as a class, to consider the solution methods from primitive to more sophisticated ones.

Assessment after lessons

Assessment should be consistent during the lesson and after the lesson. An example of this is given below. The evaluation is carried out according to students' ways of mathematical thinking, which were found on their worksheets.

Example—Manipulating on mathematical expressions: seventh grade (Ohta 1990)

The teacher used a "number game" to introduce manipulation on mathematical expressions in a two-hour lesson as follows:

Double a number that each student selects, add 2, and multiply the result by 5.

Students were then asked to find the first number when the last number was given. One part of the lesson with the teacher's analysis is as follows:

One student grasped the structure by using a figure, and another student (S1) grasped a more abstract explanation, shown by the student's own explanation as follows:

S1: It is represented as $(x \times 2 + 2) \times 5$.

Skip + 2, and multiply.

It becomes $x \times 2 \times 5$, then $x \times 10$.

There remains 2×5, then 10.

Therefore, subtract 10 and divide by 10.

To this explanation, some students nodded their heads yes, but most students put their heads a little to one side (indicating no). In general, there was no atmosphere indicating that students understood. Another student (S2) stood up to explain as follows:

S2: Suppose the first number is 2. (But he also used $\bigcirc$ and $\circ$ on the chalkboard.)

$(\bigcirc) \times 2 = \bigcirc\ \bigcirc$
If I add 2 to this, as $\circ\ \circ$ $\Big\}\times 5$, therefore

ten times $\bigcirc$

$\circ\ \circ\ \circ$
$\circ\ \circ\ \circ$
$\circ\ \circ\ \circ$
$\circ$

saying aloud in numbers, four, six, thirty. From this number, I can find the answer by subtracting 10 and dividing by 10.

S: Excellent! We understand well!

S2 supposed the first number to be 2. Although depending on this number at first, gradually S2 became detached from it and gained insight into the relation. This can be seen from the fact that S2 said "ten times" without figures. After students gained insight into the structure in this way, many students recognized the explanation by using letters.

After about one month, the teacher used the same type of problem on the semester examination as follows:

Number game:

1. Choose a positive number and add 3.
2. Multiply the result of (1) by 2.
3. Subtract 3 from the result of (2).
4. Multiply the result of (3) by 5.

If the result of (4) is 155, what is the original number?
How did you find it? Explain how to find it.

The teacher analyzed students' explanations and found seven types of meaningful responses concerning the use of letters as follows:

- Uses a literal expression (roughly) and tries to explain by transformation
- Explains by using numbers but does not depend on the numbers from the viewpoint of content
- Explains by depending on numbers but cannot detach from numbers
- Finds a relation inductively
- Explains by figures
- Explains by language
- Finds by the reverse process

The teacher evaluated each student according to these categories. Usually, it is difficult to carry out this type of analysis on a semester examination, since there is too little time. But if it is carried out, the result is useful not only for assigning a grade but also for obtaining instructional feedback.

CONCLUSION

In Japan, assessment in classroom teaching cannot be considered apart from classroom lessons. Japanese mathematics lessons involve whole-class instruction, have several stages, and respect the importance of lesson plans and lesson records.

In the main, classroom lessons using students' ways of thinking and the importance of formative evaluation are appreciated by teachers in general, but many also tend to depend on summative evaluation using paper-and-pencil tests. This is due primarily to constraints placed on evaluation by the dominant role of the entrance examinations in Japanese education.

At the same time, classroom practice and assessment that use students' natural ways of thinking are used by teachers. It is commonly accepted that even if a problem has only one solution, there may be several ways to find the solution. In this situation, the observation of students during lessons and analyses of students' worksheets after the lessons are important vehicles for carrying out assessment. Indeed, this is the starting point to using students' different ways of thinking.

REFERENCES

Becker, Jerry P., and others. *Mathematics Teaching in Japanese Elementary and Secondary Schools—a Report of the ICTM Japan Mathematics Delegation.* Columbus, Ohio: ERIC/SMEAC Clearinghouse (ED 308 070), 1989. (Also available from the author.)

Becker, Jerry P., and Tatsuro Miwa. *Proceedings of the U.S.–Japan Seminar on Mathematical Problem Solving.* Columbus, Ohio: ERIC/SMEAC Clearinghouse (ED 304 315), 1987.

Becker, Jerry P., Edward A. Silver, Mary Grace Kantowski, Kenneth A. Travers, and James W. Wilson. "Some Observations of Mathematics Teaching in Japanese Elementary and Junior High Schools." *Arithmetic Teacher* 38 (October 1990): 12–22.

Handa, Susumu. "When Do the Long-Hand and Short-Hand Overlap?" In *Let's Practice This Type of Lessons on Occasion,* edited by Yoshishige Sugiyama, pp. 90–126. Tokyo: Tokyo Shoseki, 1992. (In Japanese)

Hashimoto, Yoshihiko. "Classroom Practice of Problem Solving in Japanese Schools." In *Proceedings of the U.S.–Japan Seminar on Mathematical Problem Solving,* edited by Jerry P. Becker and Tatsuro Miwa, pp. 94–112. Columbus, Ohio: ERIC/SMEAC Clearinghouse (ED 304 315), 1989.

Matsubara, Genichi, ed. *Classroom Lessons That Pupils Think Deeply—Mathematics,* pp. 110–11. Tokyo: Tokyo Shoseki, 1987. (In Japanese)

Miwa, Tatsuro. "A Comparative Study on Classroom Practices with a Common Topic between Japan and the U.S." In *The Teaching of Mathematical Problem Solving in Japan and the U.S.,* edited by Tatsuro Miwa, pp. 135–71. Tokyo: Toyokan Shyuppanshya, 1992. (In Japanese)

Ohta, Shinya. "Cognitive Development of a Letter Formula." *Journal of Japan Society of Mathematical Education* 72 (1990): 242–51. (In Japanese)

Stigler, James W. "Research into Practice: The Use of Verbal Explanation in Japanese and American Classrooms." *Arithmetic Teacher* 36 (October 1988): 27–29.

Stigler, James W., and Harold W. Stevenson. "How Asian Teachers Polish Each Lesson to Perfection." *American Educator* [American Federation of Teachers] (Spring 1991): 12–20, 43–47.

Yoshida, Minoru. "A Study on the Features of Japanese Classroom Practices Based on Lessons with a Common Topic." In *The Teaching of Mathematical Problem Solving in Japan and the U.S.,* edited by Tatsuro Miwa, pp. 188–221. Tokyo: Toyokan Shyuppanshya, 1992. (In Japanese)

6

Tests Aren't All Bad
An Attempt to Change the Face of Written Tests in Primary School Mathematics Instruction

Marja van den Heuvel-Panhuizen
Koeno Gravemeijer

B EING a good teacher requires continual information on the progress of one's students; that is, what have they learned in their classroom lessons and what can be expected from them in the future? Although this information should be collected as efficiently as possible, teachers have much more to do than to investigate children.

DRAWBACKS OF THE USUAL TESTS

As long as there have been class-administered written tests, objections have been raised against this way of gathering information on children's learning progress. The general objection is that the usual class-administered written tests reveal only the results and tell nothing about the children's strategies. The consequence of this lack of information on children's strategies is not only that wrong conclusions are likely to be drawn about the children's performance but also that too little information is obtained about the progress of instruction. For example, nothing is learned about the children's informal knowledge and solution methods (see, for example, Ginsburg [1975]). Another consequence is that such tests make it difficult to diagnose the children's mathematical difficulties: any error analysis that depends solely on the results can never suffice to discover the children's problems and misconceptions. A final drawback of these tests is that they are too narrow with regard to both the subject matter and the children. Usually the tests are restricted to subject matter that can easily be tested and do not allow the children to show what they are able to accomplish. In other words, a lack of certain abilities may very well be balanced by other abilities not revealed by the tests.

This article was published earlier in *Realistic Mathematics Education in Primary School: On the Occasion of the Opening of the Freudenthal Institute,* edited by Leen Streefland (Utrecht, Netherlands: CD-B Press, Utrecht University, 1991).

ALTERNATIVES

As an alternative to eliminate the failings of written tests, individual observation and interviews are often considered to be the only valid way to investigate children's knowledge and abilities. This, however, need not be the case. Among the evaluation methods available to the teacher, written tests should not be wholly dismissed. They do allow the teacher to screen a whole class and should, therefore, be altered rather than rejected completely. Because there has been a tendency to look outside the sphere of formal tests for alternative means of evaluation (e.g., observations and interviews), less attention has been paid—certainly on the primary school level—to alternatives within the range of class-administered written tests. (With regard to secondary school education, the story has been different; see Lange's [1987] research on the "two-stage task," the "take-home task," and the "essay task.")

In this article we shall investigate written tests on the primary school level that illustrate a different perspective. We present a variety of possibilities to refute such objections as the lack of information about children's strategies. Some examples will be taken from batteries of tests developed in the Netherlands in the MORE project[1] (Heuvel-Panhuizen and Gravemeijer 1990). With one exception, the examples are from the third grade (eight-year-olds). The tests are meant to be administered by the teacher to the whole class. Each student gets a test booklet, and the teacher orally gives the instructions for each item.

TESTS WITH A DIFFERENT FACE

The development of these tests was closely related to the development of a new approach for teaching mathematics and was strongly influenced by the ideas of Freudenthal (1978, 1983).[2] In accordance with this new didactics we have tried to produce tests that provide a *maximum of information* about children's knowledge and abilities, covering the whole *breadth and depth of the mathematical area.* The major concern is that *children are allowed to show what they are able to do.* At the same time, the tests must be as easy to administer as the usual written tests that consist solely of numerical problems. In other words, the tests are to be applied in the classroom with a minimum of explanation; that is, with no extensive oral or written instructions that would result in overemphasizing listening or reading comprehension rather than an

---

1. The MORE project is a collective research project of the Research Group OW&OC and the Department of Educational Research, both of the State University of Utrecht. It is supported by a grant from the Dutch Foundation for Educational Research (SVO-6010). The project is an investigation into the implementation and effects of a new didactics when compared with the traditional type.

2. Treffers (1987) made an a posteriori description of the theoretical framework of realistic mathematics instruction. He gave the following characteristics of the realistic approach: the phenomenological exploration by means of contexts that serve both as source and as field of application, the broad attention paid to models that have a vertical bridging function, the use of children's own productions and constructions, the interactive character of the learning process, and the intertwining of learning strands (see also Treffers and Goffree [1985]).

understanding of mathematics. For this reason, we have looked for tasks understandable to each child that require no additional information beyond the minimum of instruction needed to get the intent across. Figures 6.1 to 6.3, which relate respectively to adding and subtracting, ratio, and measurement and geometry, illustrate the kind of items that have resulted. Little text is used; rather, pictures that are self-explanatory and related to meaningful situations convey the problem to be solved.

Figure 6.1, for example, shows two children comparing their heights. (The reproductions of the test items have been reduced here; the true size is 12 cm by 17 cm.) The obvious question is, How much is the difference? The question for the item in figure 6.2 is, How many peppermints are in the small roll? In figure 6.3, the word *over* means *left,* and the question is, How many packets of chocolate are left when the box has been packed?

These real-life contexts not only help the children immediately grasp the situation of the items but also offer the opportunity to test the children's abilities while avoiding obstructions caused by formal notation. In this way, concepts that have not yet been formally taught can be probed to provide important information for instruction. Even if this information is not used to develop new educational activities, the understanding gained about children's informal knowledge and solutions can be used to refine teaching on the basis of the children's prior knowledge. The "ratio" and "measurement and geometry" tasks in figures 6.2 and 6.3 are good examples of items that anticipate instruction, as does the item in figure 6.1. By means of the context, children may be able to answer the item even before they can do numerical problems like $145 - 138 = \square$.

Fig. 6.1

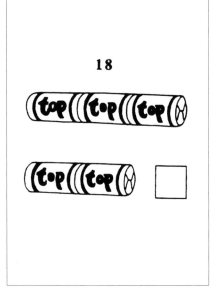

Fig. 6.2

Moreover, these items show how contexts can contribute to the understanding of test problems by suggesting clever strategies that make the items easier to solve. The picture of the two children comparing their heights elicits a complementary strategy for the subtracting task. In addition to demonstrating what these tests look like, the three examples also indicate how they can cover a wide mathematical range (addition and subtraction, ratio, measurement, and geometry). The following examples show that extensions are possible not only in the breadth of the subject matter but also in complexity. The test items need not remain restricted to flat one-step tasks that can be performed directly, without any preceding analysis. They can also investigate more complex problem-solving activities. In figure 6.4 an example of a one-step task is shown, in which a youth brass band buys four second-hand trumpets at a cost of 210 guilders each. The teacher reads, "The youth brass band buys four trumpets. They are second-hand and cost 210 guilders each. What will be the total cost? Write your answer in the box." The question of how much they cost altogether suggests adding or perhaps multiplying. Compared to this, the item in figure 6.5 offers considerably more depth. Here the question is, How many packets of chocolate are needed for 81 children? In this example, the needed arithmetical operation is not obvious; moreover, even a correct calculation of 81 ÷ 6 does not directly yield the appropriate answer. In the case of figure 6.6, where the children are asked to estimate the height of the neon letters on the building, data needed for the calculation are lacking. To arrive at a solution, the children must appeal to their own knowledge of measures (see figs. 6.22

.... **pakken over**

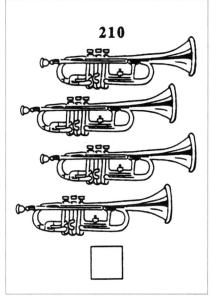

Fig. 6.3 Fig. 6.4

through 6.24 for similar types of items). Still another example of a higher-order problem is the item about the small train in figure 6.7. The question is, If the short ride takes 10 minutes, how much time will the long trajectory take? Although apparently a simple question, no indication is given about the operation to be used. On the contrary, a good analysis of the problem is required.

81 kinderen

Fig. 6.5

Another feature of the last item is a certain built-in stratification. The way the item is presented allows solutions on several levels. In addition to the use of motivating and supporting contexts, the item supplies opportunities for the children to show what they are able to do. For instance, children who do not yet have any idea of what happens to the perimeter when the length and width are doubled can, indeed, operate on a lower level and derive a solution.

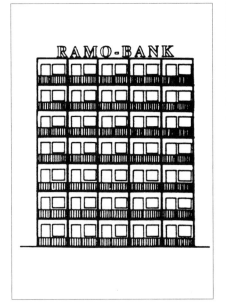

Fig. 6.6

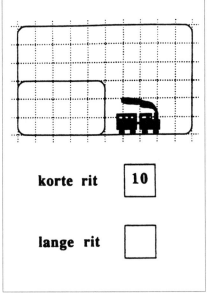

korte rit   `10`

lange rit

Fig. 6.7

MORE INFORMATION

To determine how written tests can yield more information on what children know, we can proceed in two ways. We can vary the presentation of the problem, or we can vary the ways in which children are to respond. We shall discuss three variants of both, starting with the question.

Choices

The first action that can be taken requires test items that allow opportunities to choose. In an interview, a subject who cannot solve one problem is given another in order to ascertain precisely what she or he knows. This can also be formalized in a test, as shown in figure 6.8. With this item, which is part of a test for grade 1, the children may choose for themselves what to buy. Since there are degrees of difficulty, their choice indicates what the children are able to do. Although preferences for a certain object may play a part, experience has shown that quite a few children make numerically equal choices on sequential test items of this kind. However, although the child shows by choice the degree at which difficult sums have been mastered, a correct answer does not

Fig. 6.8

disclose any insight into the child's knowledge of properties of operations and ability to apply them.

Auxiliary Sums

A way to investigate children's knowledge of operations and ability to apply them is by using auxiliary sums. Once the children are accustomed to the fact that one sum has already been made, the operation may look easy, even though these are sums with big numbers that have not yet appeared in the lesson. However, it is easy only if the children have insight into the properties of numbers and operations. This is evident in the child's work shown in figure 6.9 but not in the work of the child shown in figure 6.10, who obviously uses no understanding of the auxiliary sum. It is revealing that later on, when computing 85 + 58 and 143 − 86 without the auxiliary sum, the child whose work is given in figure 6.9 makes mistakes. This more or less confirms the hypothesis that properties of operations were used in the first case.

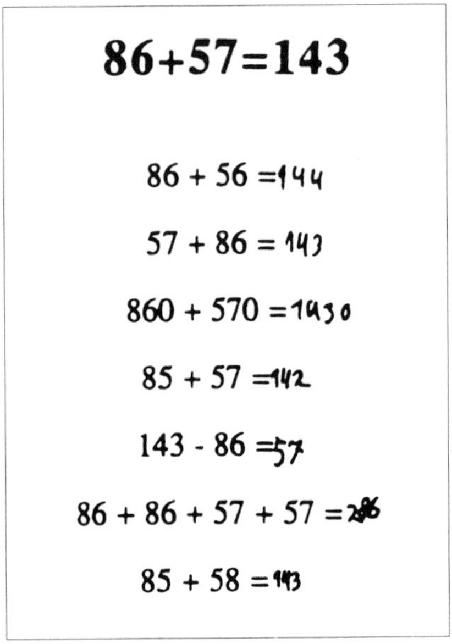

86+57=143	**86+57=143**
86 + 56 =144	86 + 56 =
57 + 86 = 143	57 + 86 =
860 + 570 =1430	860 + 570 =
85 + 57 =142	85 + 57 = 137
143 - 86 =57	143 - 86 =
86 + 86 + 57 + 57 =286	86 + 86 + 57 + 57 =
85 + 58 =143	85 + 58 =

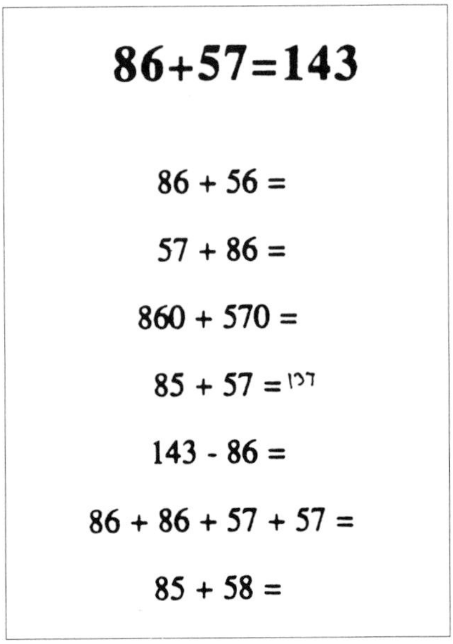

Fig. 6.9   Fig. 6.10

Change of Presentation

The last example illustrates how tests can be made more informative by presenting the same problem in different ways. In addition to presenting an item with and without an auxiliary sum, the item can be presented with and without a context. Figures 6.11 and 6.12 show examples. The problem using the context of 25 bags with 6 stamps each is easier for some children than just computing 25×6. There are, however, children for whom it is just the other way around. One child answered the bare problem at the end of the test correctly but failed the context problem. Was the item shown in figure 6.13, which was placed in between, a push in the right direction?

More Than One Correct Answer

In order to make tests more informative, teachers must consider not only the presentation of the problem but also the answer.

The first step is to abandon the erroneous idea associated with written tests that each item has a unique correct answer. By administering items with several correct answers, teachers give children the latitude to think up solutions. This not only provides opportunities for children to show what they know but at the same time gives teachers more information from tests, thanks to children's answers depicting their own level. Consider in figures 6.14 through 6.16, for

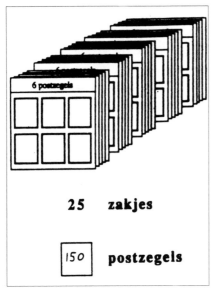

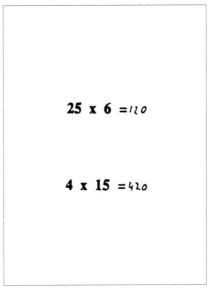

Fig. 6.11

Fig. 6.12

instance, the diversity shown in solving the problem of dividing three bars of chocolate among four children (see also Streefland [1987]).

A great variety of answers is also possible on an item about a family— mother, father, and two children—who paid a total of 50 guilders to attend the circus. The item asks children to specify the admission charges. The answers

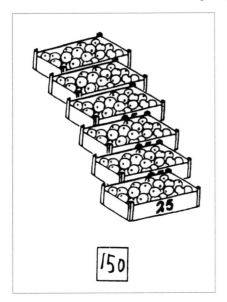

Fig. 6.13

Fig. 6.14

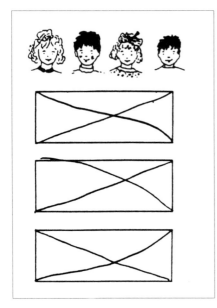

Figure 6.15   Figure 6.16

given by the children are many. Some children state that three persons pay 10 guilders and one pays 20. Other children distinguish between adults and children and respond 13 and 12, or 15 and 10 guilders, respectively. Still others evenly divide the total among the four persons to get 12 1/2 or 12.50, even though decimal and other fractions have not yet been taught.

Children's Own Productions

The best way to uncover what children are able to do is by eliciting their own productions, such as when children are asked to think up their own problems rather than to solve a given one (see also Brink [1987] and Streefland [1990]). A simple way to estimate the scope of children's abilities is to have them produce both an easy and a difficult problem. For instance, the children are asked to think up as many problems with the result 100 as they can. This can reveal not only a variety of numbers and kinds of problems but also different ways children generate problems. Some children record only isolated problems (fig. 6.17), whereas others proceed systematically by always having the first term changing by one unit (fig. 6.18) or by applying commutativity.

Besides bare problems, other types of problems lend themselves to children's own productions. One example is the item in figure 6.19, where the children are asked to make a plan for a children's party where only the starting time and activities are given. The only determined activity is forty-five minutes for the film; the children must decide how long each of the other activities should take. Like most open tasks, this one requires a great many observation points; that is,

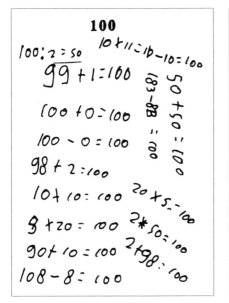

Fig. 6.17

Fig. 6.18

the times to be allocated must be decided and durations must be in tune with the activities. Finally, the digital notation of time must at least be understood.

How tests can reveal students' knowledge is shown by a test for grade 1 students. The test was administered after three weeks in the first grade and presented to a panel consisting of teachers in primary schools, counselors, teacher trainers, and researchers. The panel members were asked to estimate the probable scores of the children. The comparison of the estimated and the actual scores revealed that the numerical knowledge and abilities of the children starting in grade 1 have been greatly underestimated and that children were much more capable than presumed (Heuvel-Panhuizen 1990). Discovering this by a class-administered written test demonstrates that tests can be an important tool for teachers and researchers gathering information about children's knowledge and abilities, even for young children.

Fig. 6.19

BETTER TESTS, BETTER INSTRUCTION

This article has argued that tests need not function only marginally in the instructional process but can be integrated into the curriculum. Tests not only can help students create clever strategies but also can help teachers gain a better understanding of what children are able to do. Finally, they help teachers to develop richer didactics. This implies reversing the usual thinking about the role of tests in innovation. Rather than thwart innovations, tests can contribute to improving education.

REFERENCES

Brink, Jan van den. "Childrens as Arithmetic Book Authors." *For the Learning of Mathematics* 7 (2) (1987): 44–47.

Freudenthal, Hans. *Weeding and Sowing, Preface to a Science of Mathematical Education.* Dordrecht, Netherlands: D. Reidel, 1978.

———. *Didactical Phenomenology of Mathematical Structures.* Dodrecht, Netherlands: D. Reidel, 1983.

Ginsburg, Herbert. "Young Children's Informal Knowledge of Mathematics." *Journal of Children's Mathematical Behavior* 1 (3) (1975): 63–156.

Heuvel-Panhuizen, Marja van den. "Realistic Arithmetic/Mathematic Instruction and Tests." In *Contexts, Free Productions, Tests and Geometry in Realistic Mathematics Education,* edited by Koeno P. E. Gravemeijer, Marja van den Heuvel, and Leen Streefland, pp. 53–78. Utrecht, Netherlands: Onderzoek Wiskundeonderwys an Onderwijs Computer Centrum, 1990.

Heuvel-Panhuizen, Marja van den, and Koeno P. E. Gravemeijer. *Reken-Wiskunde Toetsen Groep 3.* Utrecht, Netherlands: OW&OC/VOU, 1990. ("Groep 3" means grade 1; the tests for grades 2 and 3 are forthcoming.)

Lange, Jan de. *Mathematics, Insight and Meaning.* Utrecht, Netherlands: OW&OC, 1987.

Streefland, Leen. "Free Productions of Fraction Monographs." In *Proceedings of the Eleventh International Conference for the Psychology of Mathematics Education,* Vol. 1, edited by Jacques C. Bergeron, Nicolas Herscovics, and Carolyn Kieran, pp. 405–10. Montreal: PME, 1987.

———. "Free Productions in Teaching and Learning Mathematics." In *Contexts, Free Productions, Tests and Geometry in Realistic Mathematics Education,* edited by Koeno P. E. Gravemeijer, Marja van den Heuvel, and Leen Streefland, pp. 33–52. Utrecht, Netherlands: OW&OC, 1990.

Treffers, A. *Three Dimensions: A Model of Goal and Theory Description in Mathematics Instruction—the Wiskobas Project.* Dordrecht, Netherlands: D. Reidel, 1987.

Treffers, A., and F. Goffree. "Rational Analysis of Realistic Mathematics Education— the Wiskobas Program." In *Proceedings of the Ninth International Conference for the Psychology of Mathematics Education,* Vol. 2, edited by Leen Streefland, pp. 97–121. Utrecht, Netherlands: OW&OC, 1985.

7

Assessing Mathematical Understanding in a Bilingual Kindergarten

James St. Clair

CLASSROOM realities affect curriculum and assessment. Age, schedule, materials, and language makeup all have an impact on what goes on. This is true in a high school geometry class as well as in kindergarten. What we do changes according to these realities, but we can also change the realities in small ways. Mathematics assessment in a bilingual kindergarten is affected by our unique classroom realities. At the same time, we adjust our "reality" to maximize our chance to do meaningful assessments. The purpose of our assessment, however, does not change. As in all classrooms, the purpose is to see what children know, how they are thinking, and how we as teachers can move thinking along.

In this article, I shall discuss three assessment techniques that work for me: teacher observations, students' recording of their work, and the sharing of these recordings. All three of these activities take into account the age and language range of my students. All three also fit naturally into a daily routine that has children actively involved and talking with each other and with the teachers.

CLASSROOM SKETCH

Here is a sketch of my "classroom reality." I should preface it by saying that I like my school and that I am given a great deal of respect and autonomy in determining my program. Still, the parameters of poverty, time, and budget give an urgent focus to what I do. There is not a lot of time or money to waste. What I teach and how I teach it determine the kindergarten experience my students will have. It's the only kindergarten year they will have. My main concern is not preparing my students for the first grade. My overriding goal is to prepare my charges to live and learn well into the twenty-first century. This "big" goal affects everything that goes on in my class.

My classroom is the English-language kindergarten of the Amigos program. The Amigos program is a two-way language-immersion program. The program is in its sixth year. Each class is balanced with English- and Spanish-speaking children. As a group they spend fully half their time in my classroom, in which English is the language we use. They spend the other half of their school time in another classroom where Spanish is used. Combined, the other teacher and I see forty different children during the school week, spending half our time with each group.

Kindergarten is a full-day program. The children arrive at 7:30 a.m. and leave at 1:45 p.m. Emily, my assistant, is with me full time. Within the last three years, budget cuts have taken away both music and science periods for my pupils. This means we have lost two valuable planning times and are now expected to "beef up" music and science in the classroom ourselves. The effect in having little leisure time is that we must use our time very efficiently.

Typically, my group will include children as old as five years eleven months and as young as four years five months. My expectation is that I'll have some children beginning to read, whereas others are learning to speak English for the first time. Likewise, it is not rare to have someone doing number operations while a classmate is practicing rote counting. Because of the range of backgrounds from which our students come, uniform goals of what children will learn are less useful than flexible, individual goals.

A developmental approach dictates that we assess what each child knows and try to focus that child on activities that maximize meaningful learning. Natural and ongoing assessment becomes crucial in this context. If we discover that a child cannot count by rote, then we will make sure there are lots of opportunities to count. Another student may count easily by rote but lack number sense. We might then design some activities involving the notion of "more or less" and reinforce the connection between number names and amounts. If at year's end we can show continuous progress in these areas, then we will have been successful.

My classroom is big and active, with a range of activities from a sticks-and-stones table to a listening center with earphones. Although we do have a complicated schedule, every day includes some time for children to choose from among a menu of choices. Mathematics is always on the menu. We have a center for games and manipulatives that changes throughout the year. Attribute blocks, Unifix cubes, geoboards, number puzzles, and board games are all rotated. Each new item is introduced to the entire class, and some time is allowed for children to just "mess around" with each new tool.

A few areas of the room allow the children to explore the physical aspects of mathematics. The water table enables them to experiment with volume. At a sand table, complete with sand timers, they can estimate time. This year, while doing lots of estimating, we filled the sand table with stones. Children spent a lot of time filling clear containers with stones and having their classmates estimate how many.

Each week also includes one or two afternoon periods devoted specifically to mathematics. These "math times" focus on a variety of activities. Some recent ones have involved (*a*) everyone practicing writing numerals on individual chalkboards and (*b*) the whole class going on a "shape walk" around the neighborhood and recording on clipboards some of the shapes found. This fall we spent some afternoons making patterns with different manipulatives and spreading out around the school to do some measuring. The variety of things we do keeps interest high and demonstrates all the different ways mathematics is part of our world. Kids seem to like coming to school.

Although English is my native language and the language of instruction in my room, I speak and understand Spanish, which allows me to communicate fully with students and parents. I listen to children's exchanges in Spanish and will use Spanish when probing a child's understanding if it brings fuller understanding. I don't consider this cheating.

OBSERVING FOR ASSESSMENT

Observing students as they work and play can furnish valuable information. This is especially true in a bilingual setting. So much learning is going on while children are actively engaged in play with each other. Second-language learners, like all children, will try out language and concepts with one another before they'll try them out with a teacher or in a large-group setting. Because there are always two adults present and because the learners are actively involved with one another and the materials, we use these "built in" opportunities to act as silent observers. We can take stock of children's understanding of concepts, their modes of thinking, and rote skills. We use this information. Both Emily and I will jot down observations on blank index cards. These dated observations go into children's portfolios. Here is a fairly typical example:

> 11/15   While making a clay pizza with meatballs, two children approached Eduardo with a clipboard and three jars of beans, asking him to estimate the number in each jar. I had filled the three jars with beans, which had 15, 25, and 50 beans in them. Eduardo estimated 14, 30, and 50. I was impressed!

This knowledge about Eduardo's estimation skill will be shared when Emily and I meet to review his portfolio before a parent conference.

We spend time jotting down lots of things. Though it is difficult to find time to share what we have observed, when we do find time we get a good idea of what different children are working on. The following anecdote is from an "activity choice" time.

11/16   Vanessa has chosen to play with a basket of counting bears and a
         basket of inch cubes. She is joined by Marisela and Jazmin, and
         they are all speaking in Spanish. Vanessa is more verbal and
         assertive than I have previously observed. She announces that
         "es una fiesta de osos" (a bear party). She begins seating bears
         on individual cubes. Marisela copies her but runs out of cubes.
         Vanessa tells her in Spanish that she needs more chairs—three to
         be exact.

This observation ended when Vanessa noticed me watching and clammed
up. But not before I noticed that she had a sense of one-to-one correspondence,
that she knew Marisela had more bears than cubes, and that she could instantly
recognize the amount three. This particular anecdote was especially important
because a few days earlier I had tried to ask Vanessa some questions while she
was working with these same materials. She smiled and let me watch, but she
wouldn't answer me. In this situation, with a shy Spanish speaker, this informal
anecdotal observation is much more meaningful than some formal task that
Vanessa might refuse to do or do wordlessly. I took time to write down the
essentials of what I had seen. After school I took time to write it out more
thoroughly to make sure I would remember what I had seen.

Not all observations contain such clear insight as these two, and a lack of
time sometimes prevents us from reviewing all the information we have
gathered. Still, a classroom setup that allows us to be watchers and our
continual practice of observing, recording what we observe, and reviewing it for
meaning furnishes many snapshots of children as learners. These very
observations can confirm what learning is going on and suggest activities for
further learning.

THE "SHARING" MEETING

On a typical morning, our twenty students might be working in eight
different areas: some in pairs, some with a teacher present, and some on their
own. As teachers, we cannot observe or be a part of everything that is going on.
There is nothing wrong with this, except that some kids might feel left out and
we miss out on things that are going on. In order for everyone to feel that he or
she has accomplished something and to communicate what has been going on
throughout the classroom, we have "sharing time" at the end of each activity-
choice time.

Five to fifteen minutes before we clean up, we warn children that activity
time is coming to a close. We ask, and sometimes insist, that children record
what they have been working on. Clipboards are available, and kids work alone
or in small groups recording what they have done. They might use numbers,
names, or even a crude graph to record the results of a board game. A group that
had been measuring each other with Unifix cubes might do a pictorial

representation of themselves and their corresponding line of cubes. Not everyone records every day, and not all recordings are shared. Likewise, some recordings go home, whereas others stay in portfolios. We even have a giant clipboard for recording structures made with the large unit blocks: simply a $2' \times 3'$ piece of plywood with a big clip to hold easel paper, it is often carried reverently to the sharing meeting by two or three young builders eager to discuss their latest design. These drawings of their own constructions show kids' sense of proportion and spatial awareness. They also preserve the structure, which allows everyone to discuss it later.

Recordings are put aside while the room is cleaned up. Then, before lunch, we have a sharing meeting. Children talk about what they've been doing. Recordings can be shown and discussed, along with paintings, collages, writings, and stories from the dramatic-play area. Children are encouraged to ask questions of one another and to comment on what is being shared. Depending on the level of fidgetiness on any given day, from three to eight children will have a chance to share something.

The sharing meeting itself is quite a learning experience. As children talk about their own projects, individual experiences become shared experiences and have an impact on other students. A recent discussion of some measuring the children had done led to a talk about ordering objects from largest to smallest. At times like this, the sharing meeting becomes a mini–math lesson. Emily and I can jot down enlightening comments that children make as well as our own observations.

At a recent sharing meeting, Kaitlin related that she was playing at the sticks-and-stones table with Yelizka and Robert. Sean asked jokingly how many stones were on the table. Kaitlin laughed, "It would take me too long to count them all!" So he asked Yelizka, who offered that there were "a lot." Robert's answer to the same question was, "Eleventy hundred." Immediately everyone in the group wanted to venture an estimate. Emily and I tried to keep track of the varying senses of number that children showed. Thus an activity in which only three students had taken part became an interesting estimation problem for all the students and an assessment opportunity for the teachers. We were able to see who used real numbers (one child squinted his eyes and said 750), who made up big-sounding numbers (the eleventy hundreds), and who was overwhelmed by the large amount (and content to say "too many").

STUDENTS' RECORDINGS AND ASSESSMENT

In kindergarten, a picture really is worth a thousand words, and the recordings that children do of their work are usually pictures with a few words as labels. Some children who have begun to write on their own will use writing, but we don't push it. We consistently use the verb *record* and let children "record" as they will.

These recordings serve many functions. They can be a record of how a child has spent his or her time, and they ensure that children will spend some time using the tools of writing. The process of recording forces children to consider what they have done and how they might represent it. Often the recording is a cooperative activity that has kids discussing how they shall express what has happened. Then, as children describe their recordings to their classmates, they develop their skills as communicators.

While the class is sharing, we can jot down children's own comments and our observations right on the recordings. The record of what they have done then becomes even richer, with kids' direct quotes and teachers' comments explaining the process. Children feel that what they have been doing is important, be it a math game, experiments with bubble making, or a satisfying game of store. These recordings can go right into students' portfolios and become part of the record of what they are doing in school.

Some examples of these artifacts follow. I have chosen examples that are related to mathematics, that are clear, and that children chose to leave in their files (if a student wishes to take home a particularly interesting recording, we try to make a copy of it before school is over). We try hard to avoid making quality judgments of recordings, since what is more important to us as teachers is what these documents tell us about a child's understanding and about what is happening in our classroom. The recording in figure 7.1, for example, shows

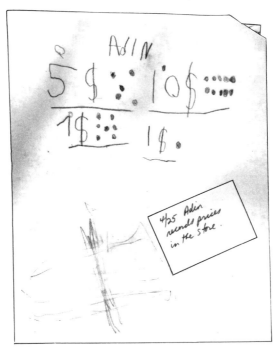

Fig. 7.1. Student recording

that this student is developing a sense of number by being able to represent numbers in different ways.

Adin had spent the entire morning in the dramatic-play area, which was being used as a store. A shy boy, this activity represented some risk-taking on his part. There was much discussion about prices before the store opened, and many trips were made to the drawing-and-writing area to make money. The recording that Adin wanted to share (fig. 7.1) took about ten minutes for him to do. In the bottom right-hand corner is my stick-on note with a date and notation. I also recall that it precipitated a lively discussion about what is expensive and what is cheap.

What Adin recorded shows that he can write numbers, knows the amounts they represent, and can present data in a clear and organized way. Given more time, he was fully prepared to try writing the names of the items that sold for these prices. What one doesn't know from this recording is that this was a youngster who was very reluctant to write when pressured to do so. If I had asked him to sit down and write the numbers he knows and what amounts they represent, I probably would not have gotten half this information and the process would have been painful and time-consuming. Instead, while I was helping other groups of children finish their work, this youngster willingly put all this information down on paper so that he could share it.

Figure 7.2 shows how evaluating a recording out of context could be misleading. Caution should be taken to see all recordings in the context in which they were made.

Jazmin had spent about fifteen minutes filling a 1-to-10 Unifix staircase with cubes, placing a counting bear on each cube, counting them for verification, and finding the corresponding numbers among a set of number caps. She had worked alone, and being a very social and verbal youngster, she was eager to share her work. I agreed she should share it and suggested she might record it so that she could share it at home as well. She agreed to record but was not quite as enthusiastic as Adin was.

The recording that she did could be misunderstood if it were evaluated all alone. One might think she saw no difference between the 1-to-10 staircase she had worked on and the brief 1-to-4 one that she drew. But Jazmin announced to everyone that her recording was "just a picture" and didn't "have everything." Moreover, she complained that she didn't have enough time either. Since her recording was imperfect, she refused to take it home. Although her recording may not be an accurate record of what she did, it did serve as a reminder of the activity Jazmin had done. Her picture also pushed us to make sure there was sufficient time for students to record properly and without pressure.

The next recording may be my favorite recording of all time. The moon table (so named because of its shape) was covered with white kidney beans and pennies. This had been someone's idea two days earlier. Yelizka and Jilcenia, inseparable pals, had made different choices earlier at the morning meeting

Fig. 7.2. Evaluation of Jazmin's recording. The recording could be misleading if looked at out of context.

when I had suggested making penny-and-bean patterns. Seeing and creating different patterns is an important part of our kindergarten mathematics curriculum, a sort of prenumeric mathematics.

Children had been making patterns with everything in the room. Pennies had been used in a sink-or-float experiment, and the beans were being planted. Someone started making penny-bean patterns, which had caught on. Emily and I were on an active campaign to have Yelizka and Jilcenia work independently of each other. Yet as I was announcing ten minutes until clean-up time, there they were—each making her own penny-bean pattern. "I expect you'll make a recording so that you can share those," I said. Yelizka, the older of the two girls, answered, "I can't draw a penny." "Hey," I responded, "a penny can be blue and a bean can be purple." "No," replied Yelizka, "Penny is PPP...purple, and Bean is BBB...blue." "Whatever...just record it," I demanded, as I scurried off to hassle someone else. Neither Yelizka nor Jilcenia was much interested in sharing their recording. They responded to the group's questions and comments, my own included, with shrugs. Although there was little discussion, the recording told a lot. Both girls can make recurring patterns and can express these patterns using different symbols. We hadn't realized that these two girls had such a complete grasp of patterns. As a result, we began having them create and duplicate increasingly complicated patterns and encouraged both of them to use a software program on patterns we had begun introducing. Beyond mathematics, we could see that Yelizka knows two consonant sounds and that she reverses some letters. The differences in the two patterns and how they were represented showed that when these two girls work together, their thinking can diverge; they are not just mimicking each other. Without the recording, I would have seen none of this. Instead, we have this wonderful snapshot of some real, cooperative mathematical work, and this snapshot can be shared with parents and colleagues.

MULTIPLE SOURCES OF INFORMATION ARE IMPORTANT

Because we are a bilingual program, verbal intercourse is at the very essence of what we do. We want our pupils to speak about what they do. Our sharing times allow for this. Fully half the pupils are learning English as a second language. Modeling language by native speakers is important, as is the opportunity for second-language learners to practice using it themselves. Yet a number of students are not confident enough to take part in so public a forum.

Much learning goes on from student to student during the activity time. A reluctance to share does not make this learning any less important for the individual learner. Rather, it adds even more importance to the teacher's observations and the student's recordings. In our particular situation, although we emphasize the communication of mathematical ideas, we are wary of making judgments based solely on verbal communications. If I had relied solely on Vanessa's nonresponses to my questions, I would have missed valuable insights gained from silently watching her work with her peers. Likewise, Jilcenia's and Yelizka's shrugs tell me nothing. Both girls are native Spanish speakers, and both are shy. If I had waited for them to tell me about their knowledge of patterns, I might still be waiting. But their drawings of their work are full of information. They might not want or be able to express this knowledge with their newly learned English, but their recordings show what they know without any spoken language. Mathematical communication is very important. In the bilingual classroom this communication takes many forms. It is important to see children in all these different situations to get the fullest picture of their mathematics concepts.

8

Assessing Measurement in the Primary Classroom

Susan Sanford

MANY mathematical situations can arise during a typical day in an elementary school classroom. By using these opportunities, teachers can ensure that mathematics will become more relevant to students' lives, that useful information will be available for assessing children's understanding of mathematics, and that mathematics can be shown to connect to other areas in the curriculum. The cooperative measurement activity discussed in this article was the result of "show and tell" in a first-grade classroom in a San Francisco public school, which led to a written measurement assessment and further development of measurement concepts.

In the first grade, measurement activities must be very concrete and at a level the children can comprehend. Piaget (1948) found that children begin to conserve length between the ages of six and nine. Before this, they rely on visual perceptual clues in making judgments about relationships involving size. Children's ability to conserve length comes from logico-mathematical knowledge that the length of an object remains constant. This is a relational concept that children must make in their heads (Kamii 1991). Children need many experiences with length and measurement to reach this level of development. First grade is a good place to give them these experiences.

Measurement is one of the ways of connecting mathematics with the environment. It involves language and terms such as *longer, the same as,* and *shorter.* It also involves estimating, matching and comparing, the understanding of relationships, and the use of pictorial representation and graphing. Measurement activities must be meaningful ones that arouse the children's curiosity and hold their interest (Kamii 1991).

For show and tell, the children are required to bring to class an object in a bag with three written clues, which are read to the class. One day Daniel brought a snakeskin, which his father had brought back from Vietnam. With the skin concealed in a bag, he read his clues:

1. It is long.
2. It was alive.
3. It does not have legs.

Three children made guesses from the clues. They guessed a toy snake, a snake, and the "shedded" skin of a snake. Daniel showed us the snakeskin. It was a *long* snakeskin, and the question of how long it was arose. The children wanted to measure it. This led naturally into a cooperative measuring activity.

We began with a discussion of standard and nonstandard measurement units. We listed things in the classroom that could be used as measurement units:

> A yardstick
> Multilinks
> Unifix cubes
> Crayons
> Feet
> Books
> Shoes
> People
> Plastic chain links
> Rulers

While listing these units of measurement on a class chart, I recorded the names of the children who made each suggestion to help me in my assessment of the children's understanding of measurement.

In cooperative groups of six (the children sit at tables of six), the children came to a consensus on the measurement unit they would use to measure the snakeskin. The discussions were mainly one- or two-word exchanges: "Shoes." "No, Unifix cubes." "No, kids!" "Okay, Unifix cubes." On reaching consensus, each group was placed on the list for its turn to measure the snakeskin. Each group measured the skin while the other children worked on other activities at their desks. Two groups chose to measure the snakeskin using children as the unit of measurement, one group chose to use shoes, one group chose to measure using plastic chain links, and one group used Unifix cubes. The snakeskin was 3 children long, 17 shoes long, 105 links long, and 11 feet, 6 1/2 inches long using Unifix cubes.

Of the groups using children, one found the snakeskin "about" 3 children long, and the other, 3 children and 1 foot long. Later as a class, we measured the snakeskin again using the tallest children in the class and the shortest. Using tall children, we found that the snakeskin was 2 1/2 children long, whereas with shorter children it was about 3 children long. We discussed this difference and the need for using a consistent unit of measurement.

The group that selected shoes quickly took off their own shoes and found they had to borrow more shoes from the children at their desks. The snakeskin was 17 shoes long. The group that chose plastic chain links put links together and carefully laid out the resulting chain along the length of the snakeskin. They then removed the extra links. The snakeskin measured 105 links long.

The group that chose Unifix cubes decided that each Unifix cube was one inch. They put twelve Unifix cubes of one color together to make one foot. They placed this next to twelve of another color along the length of the skin. Using this method, they found that the snakeskin was 15 feet, 3 inches long. I asked how they knew one Unifix cube equaled one inch. Davey had just assumed it, and so had his group. Davey suggested that we measure a Unifix cube and went to get a ruler. One Unifix cube is less than an inch. I suggested that they find how many Unifix cubes equals one foot. Davey quickly did this and found that 18 Unifix cubes equals one foot. The group realized they had to remeasure the snake. This time it was 11 feet, 6 1/2 inches.

Each child then recorded the results of his or her group's findings on a blank piece of paper. The results became a useful way to assess each child's understanding of the measurement process and his or her ability to record that information. Because this was a cooperative activity, the children were encouraged to work together. Although group work often encouraged children to produce similar papers with the same mistakes, speaking with them individually easily determined their actual understanding and contribution.

The children who used people as their unit of measurement had the most accurate written work, as seen in figure 8.1.

Fig. 8.1. David's paper shows an understanding of the measurement activity. His recording, however, by failing to line up the children head to toe, does not show the accuracy needed in measurement.

Several children who measured with shoes called them (the shoes) "inches" when writing their findings, as in figure 8.2. This points out how early social (conventional) knowledge about measurement, with such terms as *feet* and *inches*, is part of children's language (Kamii 1991). But even though they may use the words, they do not understand them. Their illustrations also did not show seventeen shoes but rather the number of shoes that fit the snakeskin they had drawn. These children may not have mastered the concept of conservation of length yet.

Fig. 8.2. Rachel calls the shoes "inches," uses invented spelling, and draws only the number of shoes that fit the snake she has drawn.

One child who had used chains as his measurement unit wrote 1005 chains instead of 105 (see fig. 8.3), suggesting the need for place-value activities.

In figure 8.4, the child understood the measurement activity and accurately recorded the findings.

The children's work can be classified into three groups.

1. The children who demonstrate an understanding of the activity and successfully record their findings, as in figure 8.4, are ready for more challenging measurement tasks.

2. The children who show partial mastery of the activity and are in transition, as in figure 8.1, need similar activities to allow them more experiences at this level.

3. The children whose work shows a lack of understanding, confusion, or unreadiness for this level of measurement activity, as in figure 8.5, need more experiences with comparing lengths of objects, placing objects in a serial arrangement, and using the terms *longer, the same as,* and *shorter.*

Fig. 8.3. Andrew's group measured the snake using plastic chain links. He understands the measurement activity but needs work with place value.

Fig. 8.4. Emma's work shows an understanding of the activity and the ability to record it with some accuracy.

These samples of the children's written work make excellent examples for their portfolios. Written work from a series of measurement activities like this one can clearly demonstrate a child's understanding of, and progress in, measurement. Assessment for this lesson took many forms: teacher observation, a class checklist to record the children's observations, the children's written

Fig. 8.5. Blair's mathematical concepts are at the readiness level. His group used Unifix cubes to measure the snakeskin. Blair's paper shows the snakeskin, 4 Unifix cubes, and some numbers. He explained that his partner said he could copy his paper, but he was unable to explain the numbers he had written.

work, class discussions, and individual interviews about the activity and about each child's written work.

I concluded from this activity that this class needs many more opportunities with measurement. Only nine of the thirty children in this class conserved length at the time of this activity. First grade is the proper time for assessing children's understanding of mathematical concepts and providing appropriate and challenging activities to help them develop a strong mathematical foundation on which to build.

REFERENCES

Kamii, Constance. "Measurement of Length: Its Development and Implications for Teaching." Paper presented at the meeting of the California Mathematics Council, December 1991, Asilomar, California.

Piaget, Jean, Barbel Inhelder, and Alina Szeminska. *The Child's Conception of Geometry.* New York: Basic Books, 1960.

9

Interviews: A Window to Students' Conceptual Knowledge of the Operations

Deann M. Huinker

THE *Curriculum and Evaluation Standards for School Mathematics* (NCTM 1989) includes a standard on Conceptions of Whole Number Operations. This standard emphasizes the importance of connecting the mathematical language and symbolism of the operations to problem situations and informal language. It also describes the importance of "operation sense"—being able to recognize real-world situations for specific operations, being aware of models for an operation, and understanding relationships among operations. Operation sense enables students to apply and use operations with meaning and flexibility. Including this standard in the document marks the importance of devoting instructional time to developing conceptual knowledge of addition, subtraction, multiplication, and division as well as assessing students' knowledge of these operations and their use of this knowledge in problem solving.

Interviews are a method of assessment that allow us to gain insight into students' conceptual knowledge and reasoning during problem solving. With paper-and-pencil tasks, students' understanding is often hidden. When a student produces an incorrect answer, it is difficult to determine whether the student lacked the necessary conceptual knowledge or made a simple error. Conversely, sometimes students get the right answer for the wrong reason. Advantages of using interviews include the opportunity to delve deeply into students' thinking and reasoning, to better determine their level of understanding, to diagnose misconceptions and missing connections, and to assess their verbal ability to communicate mathematical knowledge. An additional benefit of using interviews occurs as students provide detailed information about what they are thinking and doing—they realize that this knowledge is valued. The main disadvantage of using interviews is time. It takes time to conduct interviews, record the results, and interpret the results. With careful planning and organization, this barrier can be removed.

GUIDELINES FOR CONDUCTING INTERVIEWS

Before the Interview

What do you want to learn about your students' mathematical knowledge? Once you select a topic to assess, think carefully about what students are to know and how this knowledge is structured. Then prepare a few key questions and tasks for the interview. Examples of interview questions and tasks to assess conceptual knowledge of the operations are described later.

How are you going to record the results of the interview? You might want to make notes of the student's responses and actions on a note card or prepare some type of annotated checklist. Written notes should be taken during the interview or as soon as possible afterward. Do not rely on memory to make an accurate record after the interview is over. Careful records will enable you to better assess patterns of thinking and will provide suggestions for instruction. Another option is to tape record the interview. The tape can be replayed later to further assess the student's knowledge and can become part of the student's portfolio to be shared with parents.

Whom do you want to interview? If you want to interview all your students, interview two or three of them each day and you will be able to gather a wealth of information on all students over the course of two to three weeks. If you want to interview only some of your students at this time, plan on interviewing the other students throughout the school year. Students like to be interviewed and take it very seriously. Be sure that the other students understand that you value their thinking and that they will be interviewed at some other time on a different task.

When and where will you conduct the interviews? If you want to hold formal interviews, set up a special time and place to conduct them. For example, while students are engaged in writing or reading, you can call a student aside to be interviewed. Interviews can also be brief and informal. For example, as the students are doing independent work, stop by a student's desk and ask him or her a few specific questions, making note of the responses. Many of the interactions between teachers and students are a type of interview. The goal with using interviews for assessment is to have a plan in mind for the questions that are asked and to record students' responses.

During the Interview

Begin the interview by asking the student to "think out loud" by explaining as much as possible what she or he is thinking or doing. Point out that this will help you understand how they think about mathematics and will enable you to better help them. Additionally, point out that you will be taking notes so that you do not forget the insightful and interesting things that the student says and does. Then pose the initial question or task.

Now observe and listen as the student completes the task or answers your question. Allow plenty of time so that the student can give thoughtful responses. Real thinking takes time. Do very little, if any, talking. The goal is to find out what the student is thinking.

When the student has completed his or her response, ask probing questions to clarify what the student did or said. Be careful not to teach, give answers, or pose leading questions and suggestions. The goal is to get the student to elicit her or his depth of understanding.

Throughout the interview, record students' statements and actions that are relevant to the goals of the interview. It is often useful to record some verbal explanations word for word. It might be valuable to note whether students used concrete objects, diagrams, written procedures, mental arithmetic, or a calculator to solve problems. You might also want to note the student's level of confidence.

After the Interview

Record any additional notes or comments about specific events from the interview. The next step is to summarize the results with respect to the interview's goals. At this point, you may find that you need to interview the student again to gain additional information or to check the consistency of his or her responses. Finally, it is time to make decisions to determine your next instructional move.

ASSESSING CONCEPTUAL KNOWLEDGE OF THE OPERATIONS

Conceptual knowledge of addition, subtraction, multiplication, and division includes connections among (1) concrete and pictorial representations, (2) real-world representations, and (3) symbolic representations, as shown in figure 9.1. An isolated piece of information is not part of one's conceptual knowledge. Only when it becomes connected to other pieces of information does it become part of one's conceptual knowledge (Hiebert 1984, 1989; Hiebert and Lefevre 1986). Since it is these connections that make students' conceptual knowledge of the operations useful for problem solving (Huinker 1990), it is these connections that we should assess.

Students should be able to translate among the different representations and use the different forms with flexibility. These representations for multiplication are illustrated in figure 9.2. Given any of the representations, a student should be able to generate the others. For example, given the picture of the three packages of doughnuts, a student should be able to pose a word problem and

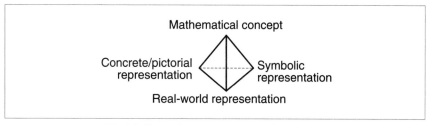

Fig. 9.1. Connections among conceptual knowledge

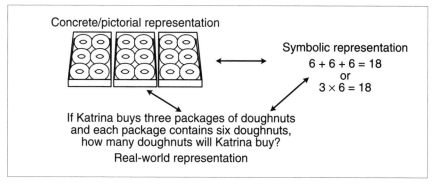

Fig. 9.2. Representations of multiplication

write a number sentence. Additionally, students should be able to communicate their knowledge using both informal and formal mathematical language.

Interview Questions and Tasks

Use the model above to guide your assessment of students' conceptual knowledge. Start with one of the representations and see if the student can generate one or both of the other representations. For each connection, some sample questions and tasks are stated below along with some possible follow-up questions.

Real-world → concrete/pictorial connection. "Use these counters (or draw a picture) to show what is happening in this word problem." Encourage the student to explain what he or she did with the counters (or drew) and why.

Real-world → symbolic connection. "Write a number sentence to show what is happening in this word problem." Ask the student to explain why he or she selected that operation and to explain the meaning of each number in the number sentence. You might want to challenge the student by stating, "Try writing another number sentence that uses a different mathematical operation." Then have the student write the number sentence and explain his or her reasoning.

Concrete/pictorial → real-world connection. "Make up a word problem about this picture (or about what you did with these counters or about the arrangement of these counters)." Ask the student to explain how the word problem relates to the picture or counters. Challenge the student to pose a different word problem for the same concrete or pictorial representation.

Concrete/pictorial → symbolic connection. "Write a number sentence that describes this picture (or shows what you did with these counters or about the arrangement of these counters)." Ask the student to explain why that operation was selected and how each number in the number sentence relates to the picture or the counters.

Symbolic → concrete/pictorial connection. "Use these counters (or draw a picture) to show me what 3×5 means." Ask the student to explain how the counters (or picture) relate to the number sentence.

Symbolic → real-world connection. "Tell me a word problem for $4 \times 23 = \Box$. How does that word problem match this number sentence? Tell me a different word problem for this same number sentence."

Additional questions and tasks to reveal students' operation sense. "What does addition (or subtraction or multiplication or division) mean? Tell me a multiplication (or addition or subtraction or division) word problem. What is the difference between addition and subtraction? How is multiplication like (or different from) addition? How is division like (or different from) subtraction? How are multiplication and division similar (or different)? When you have a word problem to solve, how do you know when you should multiply (or add or subtract or divide) the numbers?"

Connections: Established or Missing

Excerpts from interviews with Clinton and Danielle are shown in figure 9.3. These interviews examined these fourth-grade students' conceptual knowledge of division. Does their knowledge of division have connections among real-world, concrete/pictorial, and symbolic representations? Are some connections missing? Are they able to communicate their knowledge?

Clinton did not exhibit any of the connections within his conceptual knowledge of division. He did realize that division had something to do with making groups, but he did not know that the groups must be equal. Danielle had a strategy or procedure for finding the answer to the computation problem presented. She skip counted by threes and kept track of the number of threes counted. However, she was unable to represent the problem concretely and unable to describe a real-world situation. Thus, Danielle did not demonstrate any connections among the representations.

These students need opportunities to represent or act out division word problems with counters and be shown how the division equation is a symbolic representation of the situation. For example, the first number is the whole amount of objects. The second number will tell either how many parts you should make or how many objects to be put into each part, and then the answer, or quotient, will tell you the other. These students should be given opportunities

Teacher:	What does division mean?
Clinton:	… I don't know.
Teacher:	What is it like maybe?
Clinton:	Putting gum or something in groups?
Teacher:	Read that for me: $21 \div 3 = \square$.
Clinton:	Twenty-one divided into three.
Teacher:	Can you use the counters and show me how to figure out the answer?
Clinton:	[Clinton made a set of twenty-one counters. Then he separated them into two groups of eight and one group of five.]
Teacher:	So what's your answer?
Clinton:	Eight.
Teacher:	How do you know it's eight?
Clinton:	Because I put eight in a group by itself.
Teacher:	Can you tell me a story problem about it?
Clinton:	There was twenty-one pieces of gum in the bag and I ate three.
Teacher:	Can you tell me what division means?
Danielle:	Um … it's something like multiplication.
Teacher:	How do you read that? $21 \div 3 = \square$.
Danielle:	Twenty-one divided by three.
Teacher:	Can you use the counters and show me how to figure out the answer or can you do it in your head?
Danielle:	It'd be seven.
Teacher:	How do you know it would be seven?
Danielle:	[Danielle counted by threes and used her fingers to keep track of the number of threes.] Three, six, nine, twelve, fifteen, eighteen, twenty-one.
Teacher:	If you were going to show someone how to figure out the answer using the counters, how would you do it?
Danielle:	I don't know.
Teacher:	Can you tell me a story problem about this?
Danielle:	Marko had twenty-one quarters, and Marketta had three quarters.

Fig. 9.3. Assessing conceptual knowledge of division

to pose division problems about their own real-life experiences and write number sentences for these situations.

CONCLUSIONS

Interviews, like windows, allow us to see students' mathematical understanding and reasoning more clearly. As you prepare to conduct interviews, think carefully about the knowledge you are trying to assess, make plans to record students' responses, and develop an organizational plan for who will be interviewed when and where. During the interview, pose questions and tasks and then observe, listen, and ask probing questions. After the interview, summarize the results and make instructional decisions. Peering into students' thinking and reasoning through occasional interviews yields a rich source of information that cannot be discovered easily in other ways.

REFERENCES

Hiebert, James. "Children's Mathematics Learning: The Struggle to Link Form and Understanding." *Elementary School Journal* 84 (1984): 496–513.

_____. "The Struggle to Link Written Symbols with Understandings: An Update." *Arithmetic Teacher* 36 (March 1989): 38–44.

Hiebert, James, and Patricia Lefevre. "Conceptual and Procedural Knowledge in Mathematics: An Introductory Analysis." In *Conceptual and Procedural Knowledge: The Case of Mathematics,* edited by James Hiebert, pp. 1–27. Hillsdale, N.J.: Lawrence Erlbaum Associates, 1986.

Huinker, DeAnn M. "Effects of Instruction Using Part-Whole Concepts with One-Step and Two-Step Word Problems in Grade Four." Doctoral dissertation, University of Michigan, 1990.

National Council of Teachers of Mathematics. *Curriculum and Evaluation Standards for School Mathematics.* Reston, Va.: The Council, 1989.

10

Learning Logs: What Are They and How Do We Use Them?

Pamela L. Carter
Pamela K. Ogle
Lynn B. Royer

THE current emphasis on restructuring education is a prime opportunity for teachers to restructure their views and uses of assessment. No longer is assessment seen as the end product but rather as a process that enables students and teachers to become more productive thinkers and problem solvers. Now, teachers can be encouraged to use what is known about the cognitive and developmental needs of learners to teach and assess the whole person.

WHAT ARE LEARNING LOGS?

Learning Logs are journals that can be used in mathematics as well as other subject areas. These collections of student-generated words, diagrams, and pictures are used consistently and systematically to examine thinking processes and conceptual understanding. Learning Log entries are a carefully planned component of the mathematics program and can be used with learners of any age.

The idea of using Learning Logs in mathematics can be introduced to the whole class by the teacher posing questions, such as "What did you learn about line symmetry?" or "How would you describe the steps you used with the base-ten blocks to regroup and solve the problem?" After eliciting verbal responses from students, the teacher models a possible response on the chalkboard or the overhead projector. Students use this model as they experiment with writing their own responses. These Learning Log entries are discussed to look for evidence of clear thinking, a logical sequence of steps, and other indications of understanding. Working with the whole class for two to three weeks helps prepare students to write their own individual Learning Log entries.

Once students are familiar with writing entries that reflect their thinking, Learning Logs can be organized in a variety of ways. A spiral notebook, a section in a loose-leaf binder, or papers stored in a two-pocket folder can be used as a Learning Log. The entries can be placed in sections for each subject area or recorded sequentially by date of entry. Students often keep logs in their desks or at work stations where they are readily accessible. In classrooms where there is consistent access to computers, Learning Logs can be created on a word-processing program and stored on individual disks. Entries are made frequently, perhaps several times a week, to predict, reflect, and summarize during mathematics instruction and practice experiences. The work is dated and saved for students and teachers to review throughout the year.

The Learning Log contains valuable data that are used in preparing for progress reports and parent-teacher conferences. The focus is on understanding and communicating mathematical concepts and processes, not on spelling or grammar. Younger students are encouraged to draw pictures and to use invented spelling for their responses.

Teachers need to review Learning Log entries several times each week to assess students' understanding of concepts. The entries provide such a wealth of insight about students' work that teachers usually want to read and record comments on all their students. Sometimes, however, teachers may choose to select and read only representative samples of student entries and use this valuable information to monitor and adjust their teaching. Comment sheets can be included at the beginning of each log where questions, suggestions, and other comments can be written. The time needed to review Learning Log entries is usually less than that spent on checking traditional paper-and-pencil practice. Also, as teachers read these entries, much information is gained about students' conceptual development and thinking processes.

Entries at the Beginning of a Lesson

Students can use Learning Logs at the beginning of a lesson to list, for example, everything they know about polygons or to name all the "tools" that can be used for measuring. Using the Learning Logs *before* a concept is presented "sets the stage" mentally for learning that is to follow. Students might be asked to make a list or to predict how an idea might be used. Entries made before a lesson allow students to access prior knowledge and experiences in preparation for what is to come next. The example in figure 10.1 and the ones that follow are Learning Log entries from third-grade students.

Entries Made during a Lesson

Throughout the lesson, students' Learning Log entries can be used to describe their thinking and use of procedures. These entries may include notes

Situation: The introduction and use of "mental math" strategies are a daily focus in this third-grade classroom. Students have already learned the mental math strategy "counting on" and applications for different situations.

Task and Student Response: At the beginning of a lesson, students are asked to "think about a counting-back strategy. Predict what it is and how you could use it."

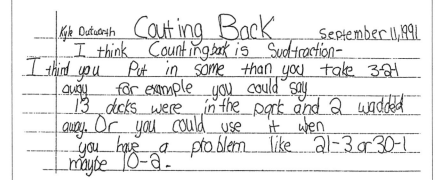

Interpretation: Kyle has used his prior knowledge about the counting-on strategy to make accurate predictions about the form and use of counting back. The reverse sequence of counting that he suggests (3-2-1) shows that he has adapted the previous strategy, counting on, to new contexts.

Feedback: Kyle was asked to share his thinking with the class so they could see how he compared counting forward with counting back.

Fig. 10.1. Kyle's Learning Log entry for counting back

about the data used to solve a problem or the steps followed in regrouping or problem solving. Although a teacher can get immediate verbal feedback while monitoring the class, the use of Learning Logs yields data to assess all students and a permanent record of each student's conceptual understanding, such as in figure 10.2.

Situation:	Elementary school students can benefit from planned instruction in using the calculator as a tool, especially for problem solving, estimation, and place value. Students who keep calculators in their desks can use them frequently during mathematics instruction and practice.
Task and Student Response:	Students are using a calculator to practice place value and are stating reasons for making their decisions. Working with a partner, they are determining which keys could be pressed to "knock out" a number from a given number in the hundreds. Yen explains how she proceeded to eliminate the 9 from the number 962.

you would have to push in 900 d know that if you push just 9 then it would take away the ones and if you push 90 it would knock out the tens so you would have to push 900 to knock out the 9.

Interpretation:	It is evident that Yen understands the value of the 9 in the hundreds place. Her use of logical reasoning and the additional detail she gives indicate she is thinking clearly about place value. A ✔ or a + can be recorded on the record sheet to indicate clear understanding.
Feedback:	Yen received positive verbal comments from the teacher. She also was asked to explain her reasoning to the class using the overhead projector to model her thinking and logical reasoning.

Fig. 10.2. Yen's Learning Log entry for using calculators to "knock out" numbers

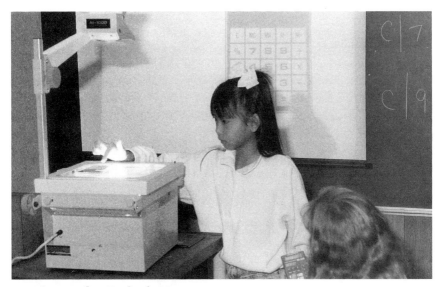

Entries at the End of a Lesson

There are many ways to use Learning Logs at the *end* of a lesson to summarize learning outcomes, such as the solution to a problem, or to evaluate the success of a particular mental math or problem-solving strategy. Learning Logs can also help teachers identify students who do not yet understand the concept or whose thinking is not complete. In figure 10.3, two students responded to the same problem in much different ways.

Situation:	Calculators are used to reinforce skip counting and problem solving.
Task and Student Responses:	Working in pairs, students were asked to record their thinking after solving this problem:
	Brandon said, "I am between 50 and 60 inches tall. You can find my height when you skip count by nines." How tall is Brandon?

Student A

We knew that Brandon. was between 50 and 60 inches tall. And we know that you count by 9 to find out how tall he is. When we were Counting by 9 and we came upon numbers like 9 and 18 but those number were under the number 50 and 60. then we kept going finely we came to a number between 50 and 60 which was 54.

Student B

My calulator skip counted for me and told me what to write down.

Interpretations:	Student A used the information given (finding a multiple of 9 between 50 and 60), his or her knowledge of skip counting on the calculator, and logical reasoning to reach a correct solution.
	Student B shows knowledge that a calculator displays an answer, but after reading this entry, the teacher is not sure the student understands the processes used. Further discussion with the student is needed. Then the Learning Log may be used again to determine if there is a more complete understanding of the process involved.

Feedback:	A written response was recorded for student A: "You used the multiples of 9 and knew you needed a number larger than 50. Good thinking!" This written response was recorded for student B: "Tell me more about what you did with the calculator to skip count. Let's talk about it."

Fig. 10.3. Learning Log entries for skip counting with a calculator

WHY USE LEARNING LOGS?

Learning Logs make connections between language and mathematical literacy. Since the use of oral and written language plays an important role in developing mathematical thinking, it is essential that students use the language of mathematics in both oral and written communication—such use of language requires that students explain what might otherwise be vague or incorrect ideas. Using the process of writing and explaining to others as cognitive rehearsal, learners are able to make sense of, and internalize, their mathematical experiences.

Another advantage of Learning Logs is to offer a mind-engaging alternative to traditional paper-and-pencil exercises. A variety of entries, which include words, pictures, and diagrams to share with others, offers a range of choices for practice and assessment.

When students learn to use multiple learning strategies, there are more opportunities to meet the diverse needs of learners in our classrooms. Also, Learning Logs actively involve all students in metacognitive behaviors by having them reflect about their ideas, solutions, and thinking processes. As students reread their entries, they often carry on a dialogue with themselves, "Now, how did I figure that out? Let's see. First, I ..." and are engaging in reflection and self-assessment.

In addition to increasing opportunities to foster student thinking, Learning Logs offer teachers "a window" through which student thinking can be observed. This is especially useful when students are working with concrete materials or pictorial representations, as in figure 10.4. Asking students to "describe the steps you used to add 428 + 346 using the base-ten blocks" provides information about conceptual understanding even after the manipulatives have been returned to the shelf.

Learning Logs also provide the opportunity for a dialogue between teachers and students through verbal and written comments. This interchange of questions and ideas offers students the chance to revise and extend their thinking. Teachers can also use this information in structuring mathematics lessons, planning mathematical experiences, and reteaching individuals or groups as appropriate.

Situation: The use of concrete materials, such as base-ten blocks or beans, is essential in developing conceptual understanding. Learning Log entries are a good way to capture these experiences on paper.

Task and Student Response: Ryan showed his thinking with a step-by-step analysis and a drawing that represented how the base-ten blocks were used to solve the problem 428 + 346. The "dangaler" (dangler) to which the student refers is the base-ten "rod" that represents the regrouping from units to tens.

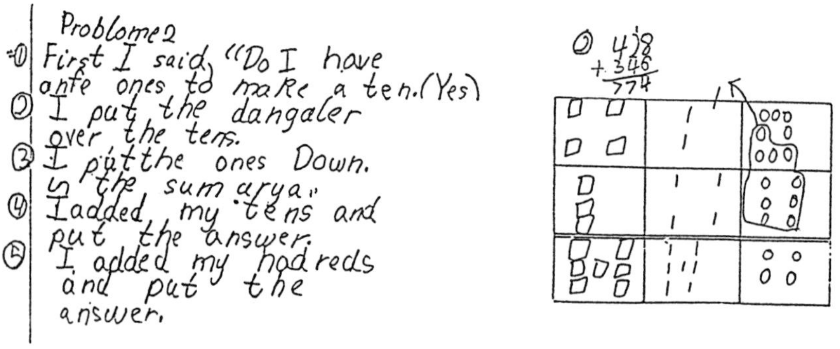

Interpretation: The teacher can see from this entry that the student understands the regrouping process and has made a clear connection between the concrete experience with base-ten materials and the symbolic representation of the problem.

Feedback: The comment "Putting the steps in order was really helpful. Good job!" was written in Ryan's Learning Log. He was also asked to assist another student in understanding the process.

Fig. 10.4. Learning Log entry for recording addition with base-ten blocks

HOW CAN LEARNING LOGS BE USED TO ASSESS MASTERY?

Learning Logs can be used throughout mathematics instruction to gather evidence of students' understanding and mastery. As teachers design a mathematics lesson, careful thought needs to be given to the criteria or evidence they will accept that students have achieved the intended outcomes. For each Learning Log entry, specific criteria, such as those listed in the earlier examples, need to be established for acceptable and intended responses. When reading Learning Logs, teachers look for clear descriptions and drawings, logical progression of thinking, and the appropriate use of terms and mathematics language. These criteria can take the form of items on a checklist, such as in figure 10.5, or be converted into points for ease of recordkeeping.

Criteria for Success

What will I accept as evidence of mastery of *the concept of equivalent fractions?*

What will the student write or say?   **What will the student do?**

• Explain the part-whole relationship.   • Use pattern blocks to show equivalence.

• Explain the idea of equal.   • Draw a picture to show equivalence.

• Reduce examples to a single fraction.

Fig. 10.5. What will I accept as evidence of ...?

Because the entries from Learning Logs are ongoing, students, teachers, and parents can observe growth over time. As areas of growth are noted and discussed, the Learning Logs become excellent tools for communicating students' progress to parents. The comment sheets mentioned previously offer a useful compilation of students' performance. Periodically, students, parents, and teachers can discuss and select Learning Log entries to be copied and included in students' portfolios. These portfolios contain a collection of each individual's best thinking and work in all subject areas for an entire school year. In some school districts, students are developing portfolios that they maintain for several years or as long as they attend school in that building. Other forms of assessment, such as student interviews and carefully designed paper-and-pencil tests, can also furnish documentation and a confirmation of understanding as noted in the Learning Logs.

LEARNING LOGS: A SUMMARY

Learning Logs are a strategic component of the mathematics program for increasing students' mathematical literacy. Included as an integral part of

mathematics lessons, Learning Logs provide the opportunity for students to learn and use the language of mathematics to demonstrate their understanding of concepts and ideas. Learning Logs also give students time for thinking and reflection. The teacher's role is to plan mathematics instruction with a variety of rich experiences at concrete, pictorial, and symbolic levels and consistently pose probing questions that elicit thoughtful and purposeful Learning Log entries. As teachers focus on criteria for success and observe their students' reflection and communication, assessing for understanding and mastery becomes part of the ongoing instructional process. The systematic and consistent use of Learning Logs as an assessment technique (1) promotes reflection and self-assessment by students and (2) provides teachers with a sequential record of their students' thinking and conceptual understanding.

11

A Day in the Life of an Elementary School Mathematics Classroom: Assessment in Action

Larry Leutzinger
Myrna Bertheau
Gary Nanke

A LTHOUGH the *Curriculum and Evaluation Standards for School Mathematics* of the National Council of Teachers of Mathematics (1989) provides a framework on which an ideal curriculum can be built, many school districts struggle with the task of implementing objectives and appropriate means of assessment that will ensure that their students learn mathematics in a way consistent with the goals outlined in the *Standards*. This article describes a model of assessment practices in a classroom in which many ideas from the *Standards* are being applied but some constraints force the continuation of traditional practices.

The following lesson from a classroom presents a guide for other teachers to follow. This classroom was selected because in many ways is it very typical of traditional classrooms throughout the country. Any teacher can recognize materials and methods he or she uses in the classroom. The teachers will feel comfortable with parts of the lesson and perhaps a bit uncertain about others. But it is hoped that the teachers will give thought to how each component fits and how the assessment procedures serve as the bonding force between teaching and learning. The crucial factor is not what gets taught but what is learned. In an actual classroom the variety of assessment procedures cited probably do not occur every day, but over a period of one week all would occur at least once.

A DAILY LESSON

Mental-Math Exercises

Each daily lesson involves several mental-math questions. These items are read to the students, who are expected to answer quickly and accurately. The teacher may allow from five to fifteen seconds for the students to write an answer. The students are to write only the answer on a record sheet. They must do all computation in their heads. These mental-math exercises assess some of the prerequisites necessary for the day's lesson and also review mental-math skills previously taught. The teacher questions students concerning the thinking strategies they used. If an interesting method of thinking is presented by a student, it can be pursued further by asking students to explain how that thinking would be used for similar exercises.

Assessment: Communication of reasoning. After completion of the daily mental-math items, selected students are asked to explain their thinking on specific exercises. Alternative ways of thinking, if present, are shared and discussed. Students can be asked to try other similar examples to verify their thinking.

Teacher: Sara, how did you think when adding 25 and 26?

Sara: I thought double 25, which is 50 and 1 more.

Teacher: Chuck, did you think that way, too?

Chuck: No. I thought 20 and 20 is 40, then 5 and 5 is 10, which is 50, and 1 is 51.

Teacher: Chuck and Sara, how would your thinking work with 45 and 46?

Emphasis should be placed on the reasoning each student uses to determine the answers and not just the answers themselves.

Periodic assessment. Every nine weeks a mental-math test is given to determine if the students have learned the mental-math skills for that quarter.

Cooperative Problem Solving

Next the students are given a problem to solve. The students work in groups of two or three and discuss possible strategies to use in solving a problem similar to this one:

Mark, Mindy, and Dan all live on the same street. Mark lives 4 blocks from school. Mindy lives 5 blocks from Mark. Dan lives 3 blocks from Mindy. How far does Dan live from school?

Periodic assessment. One student writes down the steps the team follows in answering the question. These written solutions are evaluated by the teacher using a holistic scoring method that emphasizes understanding, selection of a proper strategy, and correct completion of the problem. In the simplest scoring method, one point can be given for each of these areas. If a relatively simple scoring procedure is used, the students as well as the teacher can use it to the benefit of all. As a part of the holistic evaluation, discussion is held with each group regarding their solution procedure. Students are asked to share their methods of solution with the class.

Creating a Weekly Problem

Once a week a mathematical fact appropriate for the concept under study is written on the chalkboard. These facts come from newspapers, magazines, and television and are often brought in by the students, for example:

A baby uses 7 000 disposable diapers by the time it is trained.

The class discussion relates to how this "fact" was obtained and whether it is an estimate. The correct use of mathematical language is stressed, with an emphasis placed on proper communication and reasoning. Usually, insights are recorded by the teacher.

The students are then asked to make up a word problem containing the information from the mathematical fact. They may need to add additional information. One student asked this question: "If a box of 10 diapers costs $2.00, how much was spent?"

Assessment: Peer evaluation. Students exchange their problems and attempt to solve another student's problem. Each student offers hints to the other to assist in the solution. After a solution is determined with the steps clearly indicated, each student assesses the other student's solution procedure. They exchange their written solutions and using a three-point scale (0 points if the problem is not understood; 1 point for understanding the problem; 2 points for understanding and setting up a proper plan for solution; 3 points for understanding, proper planning, and following through to solution), evaluate the other's work. Obviously, students need to be trained in this assessment procedure. But in the long run, it is far better to have students assess one anothers' problem-solving skills than to have the teacher check each student's work. Not only is much time saved for the teacher, but the students learn much more about what is involved in the problem-solving process.

Concept Development

Most of the class period is devoted to developing the concept of the lesson. A typical lesson plan for this class is shown in figure 11.1. Notice that specific activities and evaluation procedures related to four of the five goals of the NCTM's *Curriculum and Evaluation Standards* (1989) are included for each objective taught.

For this lesson the class is organized into groups of four (see fig. 11.2). Each member is either a leader, a manager, a recorder, or a reporter. Since two of the groups have only three members, the roles of leader and manager are handled by the same person. The leader collects materials and supplies and pays close

---

Lesson Plan

Strands: Addition, subtraction, multiplication, division, place value, numeration, fractions, decimals, time, money, measurement, geometry

Problem-solving goals: Students will use manipulatives to determine all possible numbers made with six blocks.
  Assessment: An observation checklist of strategies will be used; any strategy used will be noted.

Reasoning goals: Students will use place-value concepts to determine if all possible numbers have been listed.
  Assessment: Each group will develop a logical argument that all possible numbers have been listed. Students' comments made while working both in small groups and in the large group will be noted. Written work will be collected from each group and placed in a portfolio for that group.

Communication goals: Students will work in groups and discuss the problem and possible solution strategies.
  Assessment: Comments from individuals will be recorded and included in individual portfolios.

Connection goals: Students will be able to make translations among concrete representations of numbers, spoken names for numbers, and the written form of numbers.
  Assessment: A connection checklist is used; each student is checked to see if he or she can make the proper translations.

Fig. 11.1. A typical lesson plan

attention to directions. The manager keeps people on task and silences disruptive students. The recorder writes down the group's comments and answers any questions. The reporter reads what the recorder writes and shares with the rest of the class what the group did. Students become responsible for their own learning by solving problems, discussing their solutions with one another, and then sharing their group's work with the whole class.

Assessment: Observation. The teacher is assessing at least three things by observation: (1) how the group functions as a whole and if each person is playing the proper role; (2) whether the group understands the problem, plans a solution, and solves the problem; and (3) the recorder's clarity in the writing assignment. The teacher may also hear or see an individual's or a group's special insights. To monitor the progress of the groups and to keep a record of task evaluation, a checklist similar to that shown in figure 11.3 may prove valuable to the teacher.

During the observation, the teacher should ask individual group members to say or write the number that is represented by the blocks and also ask the

Manager: Read the directions to your group and have each member do his or her part. When you finish the whole task, raise your hand.

Unit leader: Bring the hundreds, tens, and ones blocks to the group. Reach into the collection of blocks and take out six blocks.

Reporter: State the number represented by the blocks.

Recorder: Write the number represented. Can any other numbers be represented with six blocks?

Manager: Have the group discuss how to determine all possible numbers.

Recorder: Keep track of all different possible numbers.

Reporter: Read the numbers as they are written. Have you listed all the numbers you can make with six blocks? How do you know?

Recorder: Record your group's explanation of how you know you have listed all possible numbers.

Fig. 11.2. An example of a group's worksheet

	Seldom	Occasionally	Often
Cooperates			
Discusses			
Uses effective strategies			
Proceeds directly to the solution			
Shares thinking with the class			

Fig. 11.3. A checklist for group work

student to model a stated or written number. The teacher should keep a record of the student's responses similar to the one given in figure 11.4.

Assessment of shared reasoning and communication. After all groups have completed their tasks, each reporter shares what the group did. The teacher can assess the clarity of the reporter's thinking. She or he takes notes regarding the effectiveness of reasoning and communication skills for each response. These notes are included in appropriate individual portfolios.

In this classroom, the roles of the students in each group change weekly and the groups change every month. Consequently, every student is a reporter or a recorder half the time and is frequently assessed in that role.

Model to language	☐
Language to model	☐
Model to symbol	☐
Language to symbol	☐
Symbol to model	☐
Symbol to language	☐

Fig. 11.4. A translation checklist

A lot of recordkeeping seems to take place in this classroom. Recall that not all needs to be done every day. Teachers who are just begining the process of performance-based assessment should keep specific records to remind them constantly of what to assess and of the needs of the students. As teachers become more adept at assessing these thinking and reasoning skills, the assessments require fewer written records. Another benefit of the written records, at least initially, is that they furnish proof that specific learning has taken place—not a small factor with the increasing emphasis on accountability.

Teachers who do not want to use cooperative groups can use many of the same assessment procedures with individual students.

Practice in Basic Facts

The last two to three minutes of the class period are spent on practice with basic facts to review previously learned skills. After a basic fact has been learned through an exploration activity not unlike that used in the foregoing example of concept development and after the students can describe when and how to use the method on specific basic facts, those facts are practiced for speed and accuracy. This drill is best done for a few minutes every day.

REFERENCE

National Council of Teachers of Mathematics. *Curriculum and Evaluation Standards for School Mathematics.* Reston, Va.: The Council, 1989.

12

Assessment: a Means to Empower Children?

Ann Anderson

W HEN assessment is perceived exclusively as the teacher's domain, students willingly wait for the teacher to judge their success or failure. When the emphasis seems to be on external judgment, learners assume that they cannot and should not be decision makers. Assessment exclusively by one judge leads children to forfeit their autonomy and self-validation. In contrast, when children continually participate in the assessment process, they learn to recognize their own expertise. As active assessors, they necessarily exercise a more autonomous and decision-making role in their learning. Consequently, instead of being used to gain power over a child, assessment empowers that child.

ASSESSMENT: STRIKING A BALANCE WITH SELF-VALIDATION

One way to involve children actively in assessment is through self-validation, a process whereby the learner assesses the task and his or her actions in it; that is, self-validation is a form of assessment that requires the student to make a systematic accounting of his or her own knowledge of mathematics. By its very nature, then, self-validation requires learners to judge their own sense making. Therefore, the usual distinction between the roles of the one assessing and the one being assessed diminishes. When teachers include children's self-validation as a form of assessment, they legitimize the role of the independent learner. Because of its inclusive and recursive nature, self-validation serves as a valuable means to enrich learning. Indeed, my own participation in a geometry enrichment project, which involved six children in fourth grade (Anderson 1987), helped me realize that children and teachers need openly to recognize and use self-validation to complement external assessment practices.

SITUATING SELF-VALIDATION

To discuss and illustrate the role of self-validation in mathematics teaching and learning, this article includes conversations that arose as children participated in tangram activities in the geometry enrichment project. To better understand those conversations, let me describe the characteristics of the participants and the nature of the task. The enrichment group involved three boys and three girls. One boy (B3) was labeled "learning disabled," and two boys (B1 and B2) and one girl (G1) were ranked as the "brighter" mathematics students in their class. The two remaining girls (G3 and G2) were ESL (English as a Second Language) students who were considered average and below average, respectively, in mathematics. The tangram activities evolved over the later weeks of the fall semester and began with the children's solving premade tangram puzzles (e.g., flipping, sliding, rotating, or patching the seven pieces of the Chinese tangram into particular configurations as shown in fig. 12.1). This experience quickly evolved into the challenge for children to create their own tangram pieces from a square and to generate suitable arrangements to serve as puzzles.

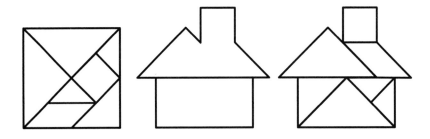

Fig. 12.1. Seven-piece Chinese tangram and examples of an empty and completed puzzle

The students' approach to these activities is important. First, the children did not consider the geometry experiences as a normal part of their mathematics class, which was characterized by an abundance of routine paper-and-pencil exercises. They appreciated the departure from the normal routine and participated with enthusiasm and commitment. Second, the children approached the tasks as generic problem-solving challenges rather than explicit lessons in mathematics. For instance, they did not approach the challenge of completing tangram puzzles by explicitly examining the mathematical properties of the pieces, but while they attempted to fill in outlines with the tangram pieces to complete a puzzle, certain mathematical properties of the pieces, the shape of the outline, and any relationships among them became apparent.

DEVELOPING SELF-VALIDATION

What the Teacher Must Do

Self-validation does not just happen, and if we are to help children develop this ability, we need to create an environment of respect, understanding, and tolerance. We must seek and consider children's opinions and encourage them to talk. We must select and construct many open-ended mathematical tasks and encourage children to assert their autonomy and creativity.

Children are strongly influenced by the traditional "teacher as sole judge" model so prevalent in our schools. As a result, when a teacher sets a task, no matter how broad, children often believe that only the teacher can determine their success or failure in completing the task. (See fig. 12.2.) To help overcome this notion, teachers who intend to encourage self-validation need to accept children's individual interpretations—at least initially—even when they seem to contradict the intent of the lesson. If we want children to believe in themselves, we must first assure them that we believe in them. Children must be encouraged to trust their own ideas more and readily to seek suggestions from, and offer suggestions to, each other. This goal can be achieved by the teacher's participating in an activity as a collaborator.

If self-validation is to develop fully in the classroom, children need experiences in which they can serve as the expert. When a solution is reached, students need to recognize that they can determine its appropriateness. When a solution is not readily achieved, they need to realize that they are a reliable source for determining the options and that no outside "authority" is needed to

---

Task: Dissect the given paper figure (i.e., square or rectangle) and use the pieces to construct a new figure. You may want to fold the paper first to mark where to cut. (Note: T is the teacher, B1 is boy 1, G2 is girl 2, etc.)

Responses

  B3: What about three-dimensional?
  G2: Could you show me what ones I've done wrong so I could fix
  them?
  T: Everything you've done is fine. There is no right or wrong.
  B3: I like making shapes that look like nothing.
  B1: You're allowed to do that?

Comment: During the first half of the enrichment project, the power of adult authority and external assessment remained strong in the students' eyes. The students feared being wrong and continually sought external judgment or guidance to determine the required response.

Fig. 12.2. Responses of students who perceive that their teacher is the authority

provide the answers. Each child's capabilities to figure out a solution must be confirmed if self-validation is to be accepted and strengthened. (See fig. 12.3.)

To optimize self-validation while doing mathematics, teachers can encourage and support independent thinking. The members of the classroom community share the authority as autonomous individuals, and each member is recognized as a capable and reliable learner. We can show children that the measure of correctness lies within the task itself and that an external authority is not always needed. Obviously, different teachers will develop the use of self-validation differently. For these students and me, participation in a nontraditional, nongraded mathematics project opened the door.

Which Tasks to Use

Although self-validation, like other types of assessment, varies with respect to both the form used and the goals served, openness seems to be a vital component. When a project is defined so that children engage their own devices to create solutions and use concrete objects to do so, judging the resulting processes and products seems to fall naturally to each child rather than to an external assessor. This was certainly so in the tangram project. To carry out open-ended tasks, children mold their project and their role in it through continuous self-validation. Questions like "Is this right?" and "Does this make sense?" are no longer addressed to the teacher but to themselves. Therefore, in such contexts self-validation is a vital component of the children's decision-making process.

If, however, tasks are constrained by such external requirements as the use of a specific algorithm, self-validation seems less likely. In conventional mathematics classrooms, many of the algorithms children use are not always well understood and often seem mysterious to them. Therefore, when specific algorithms are required, an external authority seems necessary for validation. It seems natural and rational to check the correctness of an adopted method with the source of the method—the teacher or the textbook. Such reliance on external sources diminishes when children develop their own computing algorithms. Therefore, mathematical tasks that permit learners to choose and use what makes sense to them ensures the essential link among ownership, comprehension, and self-validation.

What a Student Needs to Do

Self-validation empowers children. When students continually use self-validation, confidence in their own abilities to do and learn mathematics increases. Consequently, they need explicitly to develop and use self-validation to make decisions about many aspects of a learning task. For instance, children need to engage in self-validation in order to decide such things as which procedures to

Task: Using the tangram sets provided, complete the figures on the accompanying cards. Try to limit yourself to one bag of tangram pieces, that is, the seven original shapes.

Responses:

G1: Do this [she places two large triangles in the top to form one large triangle] and try to get a square out of the rest of the pieces.

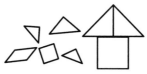

T: This is like a rectangle. [G1 seems to ignore the teacher's observation; the teacher encourages her strategy and watches as she works on the square.]

G1: Hey I did it! I just need to put it on.... Oh no, it wouldn't work 'cause it's not exactly square.

T: And that was such a good idea. I see what you mean. It would overlap.

G1 makes a couple of turns of the square to try and fit it into the remaining uncovered space. She then removes a piece; she continues to try other possibilities.

G1: I can't do it; (to G2) do you want to help me?

Comment: The materials and the process used during tangram activities supported self-validation, since children were left to their own devices to figure out the puzzles. Here, G1 used the visual cues of a fit to assess the success, failure, and validity of her "moves." Note that G1 calls on others for assistance but does not elicit the correctness of an answer. Judging whether a tangram puzzle has been correctly finished is as simple and as complex as completing it in the first place. (*Note:* On further reflection, we realize that the lower portion of the arrow is a square. The child and the teacher were both unaware of this at the time.)

Fig. 12.3. Responses of a student developing self-validation through feedback inherent in the materials

use, when to consider a solution complete, what mathematical properties and ideas are inherent in the task, and what constitutes a correct solution.

Complete tasks with minimal direction. When teachers provide minimal direction regarding a suitable procedure, children are encouraged to generate their own solutions rather than try to match the teacher's intentions. To motivate students to generate their own solutions, openness and risk taking must be common characteristics of normal classroom activity. As children begin to experiment and explore, they accept the responsibility to judge each solution path. Consequently, they more readily scrutinize their own work and that of their peers. Children, therefore, use self-validation to assess their chosen procedures regularly and to decide whether to continue with existing methods or to change.

Use personal and public criteria. Self-validation often involves the use of both personal and public criteria. Teachers need not fear that children will abuse the freedoms of self-assessment, since learners seldom make judgments solely on personal criteria. Rather, children use fairly stringent criteria in order to make very responsible decisions. (See fig. 12.4.)

Task: Create your own tangram sets. Use the given square to create tangram pieces and then use the pieces to generate cards to accompany the pieces.

Response: G1 decided to use the original tangram pieces and some of the commercial cards that accompany them to constitute one of her sets of tangrams. She traced the plastic tangram pieces onto cardboard; she used the plastic pieces to complete the commercial card and then arranged her cardboard pieces similarly onto paper to trace the outline and create her own card.

Comment: When G1 decided to copy premade sets, she chose the method (use original tangram pieces) and self-imposed the criteria (complete commercial cards). On several occasions she failed to complete a chosen card, even though she spent considerable time and energy trying to do so. As a result, she did not trace the outline and delayed the completion of her set until she was successful. Thus children can and do impose responsible and stringent criteria on self-validated tasks.

Fig. 12.4. Description of a student's adhering to stringent criteria

Decide when a solution is complete. When no specific criteria for completing the task have been discussed at the outset, some children are able to determine what should constitute an acceptable solution or an appropriate product. Even though the framework of most tasks or the underlying concepts involved furnish minimum criteria for completion, children develop and use reliable and stringent criteria for self-validation that usually expand on the implied ones. (See fig. 12.5.) In turn, when children use self-generated criteria to judge the suitability of their products, they further strengthen their abilities to self-validate.

Task: Create your own tangram sets. Use the given square to create tangram pieces and then use the pieces to generate cards to accompany the pieces.

Response: B3 had finished one card using ten or more irregular-shaped tangram pieces and was designing a second card. G1 had completed a set of cards by filling in a premade card with traditional tangram pieces and replicating the outline on colored paper.

B3: This is all I'll do with these pieces 'cause the first one is really hard.
T: Okay.
G1: I've finished another set. I think I'll make an easier one.

The teacher finds two cards on the table as the class clears away supplies.

G1: They're dumb. The shapes didn't fit. I had too many shapes. Let's throw them away.

Comment: Children make reliable and independent decisions. They expand the basic criteria underlying the activity (e.g., in a satisfactory card tangram pieces fit without overlapping) to include criteria of difficulty and aesthetics (signaled by words like *easy, hard,* and *dumb*) to judge their products.

Fig. 12.5. Responses of students developing self-validation through product assessment

Whenever children assess their own work, they capably, and often unconsciously, set valid criteria. To enhance such abilities, we need to encourage children to consciously and regularly share and discuss those responsible judgments with peers and a teacher.

Reflect on mathematical properties and ideas inherent in the task. When tasks encourage children to do and reflect on mathematics, they necessarily judge the appropriateness of the mathematical ideas involved. In the enrichment project, I chose tangram activities to have the children experience these types of mathematical problems and to explore the underlying geometric properties of the shapes involved. I wanted the children to be able to complete tangram puzzles and to recognize the properties of the pieces or outlines, which supported success in tangram activities. For the most part, the students seemed to do so, although they seldom verbalized their awareness of the mathematical knowledge their actions implied.

When the onus is on children to determine and define the mathematical task, they do notice and explore mathematical properties.When children rely on their own authority to check their progress, the mathematical sense is more relevant for them.

Use concrete materials to receive immediate feedback. When children receive direct and immediate feedback from mathematical tasks, they readily determine the appropriateness of a solution themselves. The visual cues inherent in concrete, manipulative materials not only alert students to success immediately but also permit self-correction when needed. Thus, using manipulatives in mathematics makes decision making more accessible to the child. The need to ask another if a solution is correct diminishes, and the use of self-validation is strengthened, since the materials furnish ample feedback and strategies for the child to determine if his or her actions on the materials result in the "right" solution.

When manipulative materials are involved in the task, self-validation becomes part of the interaction between the learner and those materials. A change in the materials caused by a child's actions can change an idea, which in turn can lead to further changes in the configuration of the materials or the child's intent. Such recursion requires and sustains self-validation.

Gain empowerment through making decisions. When assessment is seen as a systematic evaluation of a student's knowledge of mathematics by both the students and the teacher, then assessment becomes a more dynamic decision-making process. With such an understanding, teacher and learner make informed decisions and judgments about a task (e.g., its correctness, its versatility, and its value) and the learner's response to it. Through self-validation, students participate in such decision making as they examine the task, their personal contributions to its solution, and the suggestions of others. Likewise, as they judge their own processes and solutions against their prior knowledge, the parameters of the task, and the human and material resources available so as to decide on further actions or clarify ideas, they are empowered as learners.

CONCLUSION

Self-validation is a means to empower children, the use of which would enrich all mathematics classrooms. Self-assessment in mathematics learning encourages sense making and autonomy. All the mathematics topics used should be appropriate for self-validation. Assessment in mathematics must build children's confidence and competence. As we look for increased achievement and motivation in our mathematics classrooms, we must acknowledge and develop self-validation as one of many ways to include authentic assessment as a key element in the learning process.

REFERENCES

Anderson, Ann. "Toward a Theory of Living Systems in Mathematics Education." Doctoral dissertation, University of Alberta, 1987.

Eves, Howard. *A Survey of Geometry.* Boston: Allyn & Bacon, 1972.

National Council of Teachers of Mathematics. *Curriculum and Evaluation Standards for School Mathematics.* Reston, Va.: The Council, 1989.

Van de Walle, John A. *Elementary School Mathematics: Teaching Developmentally.* White Plains, N.Y.: Longman, 1990.

13

Assessing Students' Performance Using Observations, Reflections, and Other Methods

Ann Beyer

STOP for just a minute and think of the first student who comes to mind. Get a mental image of that student—with whom she or he likes to work, how active or motivated he or she is, how well she or he likes mathematics. Now take a sheet of paper and divide it in half. On the left side, list observations that come readily to mind about the student's performance in mathematics and her or his attitude about mathematics. Once the list runs dry, reflect again. On the right side, list the sources of information you used to make those judgments about the student's performance and attitude.

Some judgments are based on assessments such as tests; others are based on daily interactions and observations. The point is simple. As educators, we frequently use informal observations to judge students' performance. Many of us also acknowledge a need to get better at using observations and reflections to monitor and record our students' progress. However, when faced with the realities of the classroom, we often lose track of valuable information that would help us make better instructional decisions. This article addresses how teachers can better manage their assessment to determine what mathematics their students know.

Assessment methods need to be carefully aligned with outcomes—what students are to learn—and performance indicators—criteria for the evidence of progress in attaining an outcome. The model shown in figure 13.1 illustrates the interrelationship among assessment, performance, and outcomes.

Outcomes: Outcomes such as those described in the NCTM *Curriculum and*

The ideas in this article come from elementary school teachers who worked on an assessment project that was funded, in part, by the 1991 Discretionary Demonstration and Exemplary Grant for the Elementary and Secondary Mathematics and Science Improvement Act of 1988. The article describes one aspect of the project, improving teacher observation of, and reflection about, student performance.

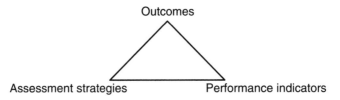

Fig. 13.1. Assessment strategies interact with outcomes and performance indicators

Evaluation Standards (1989) as well as state and local standards and objectives define important mathematical concepts and processes; that is, What is to be learned?

Performance indicators: Performance indicators answer two basic questions: What does the concept or process look like at various developmental levels? What are my criteria (i.e., what defines acceptable performance)? Performance indicators help teachers determine students' levels of performance and attainment. These indicators establish benchmarks against which students' performance can be compared. This comparison helps to determine the student's (or class's) strengths and weaknesses and, subsequently, to plan for improved instruction.

Assessment strategies: The NCTM *Curriculum and Evaluation Standards* (1989) cites the need to change the processes and methods by which student information is collected. Much has been said about the need to use alternative assessment techniques (e.g., student writing, portfolios, student self-evaluation, interviews, performance tasks). Observation, like these other techniques, addresses the question, How are you going to collect information about student performance?

GENERATING AND USING PERFORMANCE INDICATORS

One aspect of improving teachers' assessment of students' performance is for teachers to have a clearer picture of what their students' performance tells them about their students' levels of understanding. Richard Stiggins (1990) of the Northwest Regional Educational Laboratory thinks of student performance as target practice (see fig. 13.2).

As educators, we want all students to be able to hit the bull's-eye consistently (i.e., to understand a concept or process). Some students do consistently hit the bull's-eye. Others have pretty good aim, consistently hitting the target but not the bull's-eye (i.e., they are developing the concept or process). A quote by Ashley Brilliant (1987), "I may not be totally perfect, but parts of me are excellent," does a good job of describing this group of students. The third group of students routinely miss the target, their arrows falling short of the mark (i.e., they do not understand the concept or process). Performance

Who frequently misses the target?	Who consistently hits the outer rings?	Who consistently hits the bull's-eye or is close to it?
Students who do not understand the concept or process	Students who are developing the concept or process	Students who can understand and can apply the concept or process

Fig. 13.2. A range in student performance can be thought of as target practice.

indicators help to specify what student performance looks like at each developmental level, to determine which students are performing at each level, and to decide what needs to be done instructionally.

Using Your Reflections to Develop Performance Indicators

A format like the one in figure 13.3 can help in generating and using performance indicators. Think of an important concept or process you recently taught. Reflect on your students' performance. Think of students falling into one of three categories:

• Students who did not understand the concept or process

• Students who were developing the concept or process (but were not quite there yet)

	Students who are not understanding the concept or process	Students who are developing the concept or process	Students who understand and are applying the concept or process
What are the performance indicators for the concept or process at the various developmental levels?			
At what level are my students performing? Which students are at each level?			
What do I need to do instructionally?			

Fig. 13.3. A format for generating performance indicators

- Students who had attained the concept or process and could apply it and communicate it

In each column of figure 13.3 use your knowledge about student performance to make a quick draft of the evidence of thinking, behaviors, or indicators that would describe students' performance at each level. Think of this as a first draft. It does not need to be perfect, it does not need to be complete, it does not need to be in parallel form, but it does need to identify the distinguishing attributes for each performance level.

For example, a second-grade teacher drafting indicators for regrouping might include "Subtracts the smaller from the larger number in the ones place rather than regrouping" to describe a student who does not understand or "Is able to model and explain regrouping with base-ten material" to indicate a student who has attained the process of regrouping.

Now, think about individual students and write their names in the appropriate categories; then answer the following questions:

- Were you able to categorize all the students in a given class?

- If not, what could you do to assess the remaining students' level of proficiency?

- What insights do the data provide that will help you plan your next steps?

- Having this information about student performance, what do you need or want to do instructionally?

- What can you do to provide extended opportunities for students who are at the beginning stages?

- What can be done to more flexibly extend and enrich the experience for students who are demonstrating the desired level of proficiency?

Using Performance Indicators and Observations to Improve Instruction

You were just introduced to one simple strategy for both generating and using performance indicators as a self-help tool for improving instruction. By clustering student performance into the three general categories, you can begin to reflect in a more meaningful way about your students' performance, thinking, and behaviors, subsequently using this information in your instructional planning. This actually becomes a three-step process of reflection and observation:

- Generating (through reflection and observation) performance indicators for important concepts or processes

- Observing and recording student thinking, behavior, and performance

• Reflecting on performance data to monitor students' performance and to improve instructional decisions

Teachers who have tried this reflective process indicated that it caused them to look at student performance and their classroom practice with new eyes. It clarified their focus, which helped them to be more analytical and reflective in their daily observations of student performance. Because they had clearly articulated their target, they were better able to ascertain what students did and did not know, to select more appropriate instructional activities, and to better monitor and communicate student progress toward desired outcomes.

Hints for Generating and Using Performance Indicators

1. It is easiest to work with generating indicators at just three levels. Teachers usually described the two extremes first (i.e., those who did not understand and those who had attained the concept) and then the middle (i.e., those who were developing the concept).

2. Performance indicators do not need to be generated for all outcomes or lessons. Choose some areas for focused instruction or ongoing emphasis. To get started, pick a limited number of important outcomes on which you spend a significant amount of time. Begin by generating and using performance indicators only in these areas.

3. Recognizing that students mature in their ability over time and at different rates, you may find it helpful to write your performance indicators as end-of-year tasks. This will help you preassess, assess, and reassess students' progress in the context of where they are and where you want them to be by the end of the school year. By comparing their performance to the indicators, you can plan instructional activities to accommodate better your students' developmental needs. It also will help you monitor their progress through the course of the year, particularly if you also keep things like work or product samples and self-evaluations in a portfolio.

4. Some general performance patterns will emerge. To get you started, figure 13.4 lists sample performance indicators for some mathematical concepts. They are intended to trigger your own thinking. Because of the age of your students or the level of the development of the concept in your grade, some of the samples may not be applicable or they may be in the wrong column. You can move them, change them, or provide other, more specific indicators that are more appropriate for your own students.

OBSERVING AND RECORDING STUDENT PERFORMANCE

As already discussed, teachers continually observe individual, small-group, or class performance to assess some aspects of students' thinking, behavior, or attitude. A common concern is finding effective yet efficient ways to record

NOT UNDERSTANDING	DEVELOPING	UNDERSTANDING AND APPLYING
UNDERSTANDING THE PROBLEM OR SITUATION		
• Does not attempt the problem • Misunderstands the problem • Routinely requires explanation of problem	• Copies the problem • Identifies key words • May misinterpret or misunderstand part of the problem • May have a sense of the answer	• Can restate or explain the problem coherently • Understands chief conditions • Eliminates unnecessary information • Identifies needed information • Has a sense of the answer
CONCEPT UNDERSTANDING (E.G., MULTIPLICATION, SYMMETRY)		
• Does not routinely model the concept correctly • Cannot explain the concept • Does not attempt problems • Does not make connections	• Demonstrates partial or satisfactory understanding • Can demonstrate and explain using a variety of modes (e.g., oral, written, objects, model, drawings, diagrams) • Is starting to make how-and-why connections • Relates concept to prior knowledge and experiences • Can create related problems • Accomplishes tasks, though with minor flaws	• Correctly applies rules or algorithm on how to manipulate symbols • Connects both how and why • Can apply the concept in new or problem situations • Can see and explain connections • Accomplishes tasks and goes beyond
MEASUREMENT (LENGTH, MASS, CAPACITY)		
• May make direct comparisons between objects • Cannot order objects according to measure • Does not distinguish differences in measurements	• Can compare and order using nonstandard units • Can estimate and measure using nonstandard units • Can estimate and measure using standard units • Can solve some related problems	• Can estimate and measure using standard units • Can select appropriate measurement units for task • Can use fractional increments to measure • Can solve related problems
ESTIMATION		
• Makes unrealistic guesses • Does not use strategies to refine estimates • Cannot model or explain the specified strategy • Cannot apply strategy even with prompts	• Refines guesses or estimates by partitioning/comparing • Can model, explain, and apply a strategy when asked • Has some strategies; others are not yet in place • Uses estimation when appropriate	• Makes realistic guesses or estimates • Refines estimates to suggest a more exact estimate • Uses estimation when appropriate • Recognizes and readily uses a variety of strategies
VERIFYING RESULTS		
• Does not review calculations, procedures • Does not recognize if answer is or isn't reasonable	• Reviews calculations, procedures • Can ascertain reasonableness if questioned	• Checks reasonableness of results • Recognizes unreasonableness
COLLECTING, ORGANIZING, AND DISPLAYING DATA		
• Makes no attempt • Cannot proceed without direction and assistance • Makes major mistakes in collecting or displaying data	• Can collect and display data, given a method to record • Has minor flaws in collecting or displaying data • Can correct errors when pointed out	• Can collect and display data in an organized manner • Accurately and appropriately labels diagrams, graphs, etc.

NOT UNDERSTANDING	DEVELOPING	UNDERSTANDING AND APPLYING
SUMMARIZING, INTERPRETING RESULTS		
• Makes no attempt to summarize or describe data • May answer simple questions related to data if prompted • Cannot communicate results in rudimentary form	• Summarizes and describes data appropriately • Can generate and answer questions related to data • Can communicate results in rudimentary form	• Draws valid conclusions and interpretations • Makes generalizations • Communicates results clearly and logically
APPLYING STRATEGIES, CONCEPTS, PROCEDURES LOGICALLY		
• Makes no attempt • Relies on others to select and apply strategies • Work is not understandable • Cannot explain work or strategy adequately • Selects inappropriate strategies • Implementation is not logical or orderly	• Uses strategy if told • Recognizes strategy • Can explain strategy • Uses a limited number of strategies • Can select a strategy but may need assistance in its implementation • Can present work in an acceptable manner	• Generates new procedures • Extends or modifies strategies • Knows or uses many strategies • Uses strategies flexibly • Knows when a strategy is applicable • Presents work logically and coherently
MATHEMATICAL COMMUNICATION		
• Has difficulty communicating ideas • Withdraws from discussions • Cannot bring thinking to conscious level • Does not use, or misuses, terms • Offers unrelated information	• Expresses ideas in rudimentary form • Can support simple explanations with models, drawings, etc. • May need some assistance or prompts in refining skills • Uses some terms appropriately	• Communicates clearly and effectively • Explains thinking process well • Can communicate ideas in several forms (orally, in writing, drawings, graphs)
MATHEMATICAL DISPOSITION (VALUES, LIKES MATHEMATICS)		
• Demonstrates anxiety or dislike of math • Withdraws or is passive during math time • Gives up easily, is easily frustrated during math • Needs frequent support, attention, feedback	• Applies self to task • Is actively involved in learning activites • Is willing to try new methods • Does what is asked but may not take initiative	• Demonstrates confidence in work • Is persistent, will try several approaches; does not give up • Is curious; demonstrates flexibility • Asks many questions
USE OF MATERIALS		
• Needs more exploration with materials • Cannot use materials without assistance • Watches to see how others are doing it before trying it on own • Does not attempt to use materials	• Generally uses materials effectively • May require occasional assistance	• Uses materials effectively and efficiently
EXTENDING THE PROBLEM, MAKING CONNECTIONS		
• Does not attempt to make extensions • Does not make connections • Cannot extend ideas to new applications • Does minimum expected	• Can recognize similar problems or applications • Makes connections	• Proposes and explores extensions • Can create parallel problems by varying conditions of original problem • Can apply ideas to new applications

Fig. 13.4. Sample performance indicators for different goals

students' performance. That is, find strategies that work for you, not overwork you. A few of the more innovative techniques suggested by teachers are described.

Performance Indicator and Recordkeeping Sheets

One technique is to make the performance indicator sheet a recordkeeping sheet as well. The top half of the sheet lists performance indicators; the bottom half is used to enter students' names. This format helps ensure that the teacher observes each student and makes a judgment about his or her performance. The example in figure 13.5 is for graphing.

NOT UNDERSTANDING	DEVELOPING	UNDERSTANDING AND APPLYING
• Cannot complete task • Makes more than two errors in graph • Draws inaccurate or incomplete graph • Cannot or does not attempt to communicate results or make interpretative statements	• Needs assistance in in constructing graph • Makes minor errors in graph • Corrects errors if pointed out • Needs assistance in communicating or interpreting results	• Constructs graph without assistance • Draws axes, increments, and labels correctly • Graphs data correctly • Can communicate results and make interpretive statements
Gary Juan Monica Linda Sammy Mike	June Mitchell Cathy D. Bo Jeremy Mark Becky Debra Avery	Miguel Sarah Denny Jacob Sheri Bob Cathy L. Micki Chi Li

Fig. 13.5. A performance indicator sheet for graphing also used as a recordkeeping sheet

Calendar or Seating Chart Models

Teachers can use either large calendars or seating charts to record student information. In each cell a student's name is written; this space is used to record individual student observations. As a tool to encourage greater teacher reflection, this model helps teachers identify, at a glance, which students they have not observed or called on frequently enough to obtain good data.

As with any model, teachers can modify this one to make it work better in their own particular situation:

1. Try writing notes about students on Post-it notes and then affixing them to the cells at a later time.

2. Develop a simple coding system (e.g., + for correct responses, – for incorrect responses, ? for student-raised question) for recording some responses. You can still write in specific notes or phrases about the behaviors or thinking observed.

3. Consider taking the student listed on a given day and doing a mathematics interview with that student. The mutual benefits are evident: students enjoy the one-to-one time with the teacher, and the teacher gains valuable insights into the

students' level of understanding. If you try this, consider including questions that ascertain how students feel about mathematics and that involve student self-evaluation.

Note Card or Notebook Models

Many people use individual student note cards in a file or a page in a notebook for each student. The idea of this system is to have a card or page on each student. A common complaint is having to flip through them to find the one you want. Many teachers find it more manageable to use this model. You need index cards, masking tape, and heavy cardstock (a clipboard is optional). Write a student's name on the bottom of each card. Tape the cards to the cardstock so that they overlap with each student's name in view. To reinforce the card, put tape on both sides of the card (see fig. 13.6). Use the cards to record thinking strategies, specific behaviors, and insights about the individual students. The observations can be limited to predetermined outcomes on which you want to collect data. Once a card is full, it can be filed for future reference (e.g., report-card information, parent conferences).

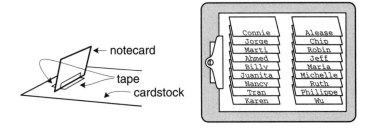

Fig. 13.6. Overlapping index cards provide a handy way to keep a record on each student in a class.

Another format is to use folders for each student. The folder holds samples of a student's work, notes, and other evidence of a student's performance. An innovative addition to this model is to keep a set of computer labels handy. Record and date notes on individual students on these labels. They can be removed from the strip and affixed to the student's folder at a later, more convenient time. If you want to be sure to collect data on each student, write names on the stickers in advance as a reminder or record of which students to observe, question, or interview.

Hints for Making Observations and Recordkeeping More Manageable

1. Limit observations to those things that cannot be more readily evaluated by other techniques (e.g., attitude toward problem solving, the selection and

implementation of a specific strategy, modeling a concept with a manipulative, ability to work effectively in a group, persistence, and concentration).

2. Ask students questions that will help you better understand their thinking, behavior, and understanding (e.g., What did you do first? Why? Can you describe your solution? Will you please explain what you are doing? What should you do next? Can you describe any patterns you see?).

3. Get in the habit of recording your observations briefly and objectively during or soon after the observation.

4. Frequently, recording observations will point out those students whom you have not observed for a while. For example, just tallying on a seating chart those students who have responded to questions will make you more aware of interesting patterns.

5. Sometimes (e.g., when observations are done over an extended period of time) you may find it helpful to code the date of the observation.

6. Play around with different recording formats until you find ones that are the most useful and user friendly.

7. Observe students in a natural classroom setting so that you can see how they respond under normal conditions. It is easier to observe students' behavior if they are working in small groups rather than alone.

8. Have an observation plan, but be flexible enough to note other significant thinking or behavior. You may find it helpful to limit the number of students or the number of things for which you are observing.

REFERENCES

Brilliant, Ashley. *I May Not Be Totally Perfect, but Parts of Me Are Excellent.* Santa Barbara, Calif.: Woodbridge Press Publishing Co., 1987.

Labinowicz, E. *Learning from Children: New Beginnings for Teaching Numerical Thinking.* Menlo Park, Calif.: Addison-Wesley Publishing Co., 1988.

National Council of Teachers of Mathematics. *Curriculum and Evaluation Standards for School Mathematics.* Reston, Va.: The Council, 1989.

Stiggins, Richard. "Assessment Training." Workshop for the Ann Arbor Public Schools, Ann Arbor, Michigan, 1990.

14

From Multiple Choice to Action and Voice:
Teachers Designing a Change in Assessment for Mathematics in Grade 1

Jane F. Schielack
Dinah Chancellor

W HEN presented with the recommendations from the *Curriculum and Evaluation Standards for School Mathematics* (National Council of Teachers of Mathematics 1989) and the teaching section of the *Professional Standards for Teaching Mathematics* (National Council of Teachers of Mathematics 1991), teachers of elementary school mathematics often respond that the suggested changes in instruction do not fit the way their districts require students to be assessed in mathematics. In direct counterpoint, most mathematics education organizations and supervisors—and many teachers—support the idea that we should assess what we think is important to teach, not determine what is taught by what is necessarily easy or popular to assess. How do we escape this seemingly endless chicken-or-egg argument?

One district managed to make a major change in its first-grade testing program from paper-and-pencil multiple-choice items at the end of each six weeks to assessment activities embedded within instruction. The plan embraced both points of view: using changes in instruction to generate a need for new assessment techniques, and then using these changes in assessment to improve instruction further. Just as in problem solving, however, the process of the change was as important as, if not more important than, the assessment product itself. During the three years of developing a more appropriate assessment plan, five distinct periods occurred: preparation, "disequilibrium," production, experimentation, and implementation.

The activity that is the subject of this chapter was produced under a grant from the Texas Higher Education Coordinating Board and the U.S. Education Department under the auspices of the Eisenhower Mathematics and Science Grants Program, Title II. Opinions, findings, and conclusions expressed herein do not necessarily reflect the position or policy of the Texas Higher Education Coordinating Board or the U.S. Education Department, and no official endorsement should be inferred.

THE PREPARATION PERIOD

The change process began with a year-long in-service project for kindergarten and first-grade teachers throughout the district. The participants came from eleven of the seventeen district campuses and served approximately 2000 of the 12 000 students enrolled. During this in-service project, the teachers met approximately twice each month to engage in active mathematics learning experiences. Between in-service meetings, they were to implement active mathematics learning in cooperative groups in their classrooms and record their experiences in a journal. As the in-service project progressed, it became apparent that several first-grade teachers were becoming very concerned about how this type of instruction was going to fit with the district's traditional six-weeks assessment plan. Just as a problem setting can generate a need for students to learn a new mathematics concept, this in-service period generated a need for teachers to investigate different means of assessment that would support the changes in instruction that they had come to believe were important.

THE "DISEQUILIBRIUM" PERIOD

During the next school year, ten teachers were chosen to work on the development of a curriculum and an appropriate accompanying assessment plan for mathematics in grade 1. Although most of these teachers had shown a high level of interest in such instructional changes as active learning and the integration of the identified content strands of patterns and relations, number, operations, geometry, measurement, and probability and statistics, they also displayed a high level of discomfort applying these changes to the entire mathematics curriculum. The teachers identified important mathematical connections among the major content strands—such as those between number and graphing, between patterns and geometry, and between patterns and basic facts—and designed a curriculum that included concepts from each major strand in each six-weeks period. The next three months of struggling to integrate the strands for each six-weeks grading period alleviated their concerns about putting this approach into practice while still covering traditional arithmetic topics.

As the curriculum became better defined, the teachers felt more and more strongly about the need for changes in assessment. Before they designed specific assessment procedures, however, they began to reevaluate their ideas about assessment in general. They identified three major purposes for assessment: (1) to communicate students' participation in mathematical thinking to parents, principals, and other instructional leaders; (2) to provide information to the teacher so that appropriate instructional decisions can be made; and (3) to provide information to the district for the evaluation of instructional programs. Although the question of assigning grades was a definite issue in the area of communicating to parents, the teachers decided that the design of the assessment should be driven by the broader purposes identified and that several grading schemes appropriate for the design would be explored during its implementation. When the mathematics supervisor consolidated the teachers' curriculum outline and assessment goals into "big ideas" from each strand to be assessed at the end of each semester, as shown in figures 14.1 and 14.2, the result of this period of "disequilibrium" was a useful summary of the teachers' questions and conclusions.

Grade 1 Mathematics
First Semester Student Report

Student: _____

Teacher: _____

School: _____

Date: _____

The information below is based on observation and assessment of your child's mathematical understanding and attitudes as he/she participated in group or individual activities throughout the first semester.

CODE: M = most of the time; S = some of the time; N = not yet

Math Concept/Attitude: Your child	Code:
shows confidence in solving problems.	
is able to solve problems in more than one way.	
identifies patterns made up of concrete objects or shapes.	
uses sorting rules to create a graph.	
creates and labels sets to 12.	
uses sets created with story problems to record addition/subtraction sentences to 12.	
uses non-standard units to estimate and compare length and mass.	
sorts and compares geometric solids and figures.	

Fig. 14.1. First semester's "big ideas" for grade 1 assessment

In examining the purposes of assessment, the teachers began to experiment with the results of assessment being displayed along a continuum, emphasizing what students *could* do rather than what they *could not* do. After much discussion, the teachers decided on the assessment categories of "most of the time" (M), "some of the time" (S), and "not yet" (N). They felt they could then move away from paper-and-pencil testing of each student individually and begin to use recorded observations of students working in groups during daily instruction as part of their assessment plan. Anecdotal comments and examples of student work also would be used to confer with parents.

THE PRODUCTION PERIOD

Once the "disequilibrium" period had been worked through and the teachers were feeling comfortable with the "big ideas" of their new curriculum, they began to identify appropriate activities for assessing students' understanding of

Grade 1 Mathematics
Second Semester Student Report

Student: _____

Teacher: _____

School: _____

Date: _____

The information below is based on observation and assessment of your child's mathematical understanding and attitudes as he/she participated in group or individual activities throughout the 2nd semester.

CODE: M = most of the time; S = some of the time; N = not yet

Math Concept/Attitude: Your child	Code:
shows confidence in solving problems.	
is able to solve problems in more than one way.	
identifies and extends patterns made up of symbols.	
interprets graphed data.	
creates and labels sets to 99.	
uses sets created with story problems to record addition/subtraction sentences.	
uses standard units to estimate and compare length, mass, and capacity.	
identifies congruent figures.	

Fig. 14.2. Second semester's "big ideas" for grade 1 assessment

these concepts. The group of ten teachers was separated into several smaller groups, each of which compiled a selection of representative assessment activities for a particular strand. As can be seen in figures 14.3–14.6, the assessment activities not only included directions for the student action involved but also suggested questions for the teacher to ask.

Some assessment activities also included suggestions for teacher observations, such as "Did the student use an organized method of adding or

---

Objective: The student will use sorting rules to create a graph.

4. Button Graph: small group

Materials: buttons, grid or graphing mat

Procedure: Give a handful of buttons to each student. Ask students to sort the buttons in some way. Have students place the sorted buttons on the grid to form a graph.

Suggested Teacher Talk: What should the title of your graph be? What can you tell me about your graph? How did you decide how to group the objects? How else might you sort these objects? How many ____ do you have? Do you have more ____ or more ____? Which group has the most? Which group has the least?

Fig. 14.3. A first-semester assessment activity for patterns and graphing

Objective: The student will sort and compare geometric solids and figures.

1. Read My Mind: small group, individual

Materials: a collection of pattern blocks, attribute blocks, color tiles (or paper representations from the Teacher Materials Center), black and white construction paper, Unifix cubes (one of each for each child)

Procedure: This is a silent game. The teacher sorts about 3/4 of the figures on the two different pieces of construction paper as children observe. Then as the teacher holds up a figure, the students predict in which group it belongs by putting their Unifix cube on either the white or the black paper.

Suggested Teacher Talk: What did I care about as I placed the figures each time? Are there any figures on the black paper that are not on the white paper? Why do you suppose I placed them where I did?

Let's try it again, only this time I will use a different sorting rule! (Sorting rules could include figures with four corners; figures with greater or fewer than four corners; figures with straight sides, figures with curved sides; red figures with four sides, figures that are *not* red and do *not* have four sides; etc.)

Fig. 14.4. A first-semester assessment activity for geometry

removing the nonstandard units of mass to balance the eraser on the scale, or did these actions appear to be random?"

In a discussion of how to translate the student responses and behaviors generated by these activities into marks of M, S, or N, the teachers felt that they should begin to compile anecdotal records while using the activities during the experimentation period. These recorded observations would then be used as part of the in-service component in the implementation period. As teachers were trained in the use of this new assessment plan, they would participate in discussions about appropriate interpretations of possible student responses.

THE EXPERIMENTATION PERIOD

Knowing they needed a basis for credibility before selling this assessment plan to other teachers, the group of ten teachers began to use the assessment activities within their own instruction and submitted the recording sheet for the first semester to the district office. Besides identifying improvements or changes that needed to be made in the activities, this experimentation period also identified some interesting patterns within the student assessments. A mark of "Not Yet" in mathematical disposition often accompanied a similar mark of

Objective: The student will create and label sets (no more than 99).

3. Fair Exchange: small group, individuals, partners

Materials: place-value mat, Dime Exchange sheet (Line Master 50, included in this packet), numeral cards (10–99), dimes and pennies in labeled containers

Procedure: Have one student draw a numeral card and place that number of pennies on the ones side of the mat. The partner will group the pennies into tens. He or she will exchange each group of pennies for dimes, which will be placed on the tens side of the mat. The first student will record the exchange on the dime exchange recording sheet, and the partner will initial the exchange. Have partners change roles after each turn.

Suggested Teacher Talk: How many pennies did you give your partner? (28) How many groups of ten have you made so far? (1 ten) How many pennies are left in the pile? (18) How do you know? Can you make another ten? If I have 36 pennies and want to trade them in for dimes and pennies, how many dimes should I get? How many pennies will I keep? How will you figure it out? Can you think of another way?

Teacher Observation: When asked to check an exchange to make sure it is fair, does the student count the dimes as tens and the pennies as ones? After making one exchange, does the student know how many pennies are left? Does the student know if another exchange is possible?

Fig. 14.5. A second-semester assessment activity for number

"Not Yet" in both the number and operation strands, the strands that have played the major part in traditional assessment. However, many of the students with "Not Yet" ratings in these areas had satisfactory ratings in the other areas that traditionally have not been assessed (geometry, measurement, graphing).

These results were examined from two perspectives: (1) evaluation of the student by the teacher and (2) evaluation of the program by the district's instructional leaders (teachers, principals, and supervisors). What did these results indicate about students' understanding of geometry, measurement, and graphing concepts compared to number and operation concepts? What did the results indicate about the kinds of assessment activities that were used for geometry, measurement, and graphing (e.g., were they more concretely oriented than those for number?)? What did the correlation between disposition, number, and operations indicate about teachers' interpretations of mathematical disposition and problem solving? Were geometry, measurement, and graphing being viewed as components of mathematical disposition and problem solving?

Objective: The student will identify congruent figures.

1. Pair Puzzles: small group, partners

Materials: paper, envelopes, scissors, a variety of real-life gometric solids (small boxes for rectangular prisms, soup or orange juice cans or pill bottles for cylinders, etc.), sets of geometric solids

Procedure: Have the children trace and cut out a face from several different solids. They should cut one of the faces twice, however, to create at least one pair of congruent figures. Next, they place their cut-out figures in an envelope and pass to a partner. The partner must take the figures from the envelope and locate two figures that are congruent.

Suggested Teacher Talk: Which solid can you trace to make a figure that looks like this one? How about this one? How many sides does this figure have? How many corners? How are these figures the same? How are they different? Explain how you know two figures are congruent.

Fig. 14.6. A second-semester assessment activity for geometry

These and other questions were used to guide the planning of the in-service portion of the implementation period.

Figure 14.7 is an example of several creative stories involving numbers that were written by students in an activity to assess the "big idea" in the operations strand—students' ability to use sets created with story problems to record addition and subtraction sentences.

Although each of these students received an M for achieving this objective most of the time, each story provided some points for further discussion. The author of the story in figure 14.7 was asked for other possible questions that could be generated from the same story. In another story, the teacher and student discussed the possibility of rewording the question to make it more clearly refer to the flowers left in the ground. To the author of a third story, the teacher posed the questions "What would make you have to take two away? How could you rewrite the story without using the words *took away*?" in order to investigate the student's understanding of the connections between subtraction and the physical actions it models. Another author was asked to describe the story with a number sentence or sentences, if possible, to assess the student's level of knowledge about whole-number division and its relationship to repeated subtraction. The teacher investigated this idea also by asking the student why the phrase "if they got the same amount" was included. The teachers were now definitely able to obtain a broader picture of each first-grade student's mathematical understandings.

Fig. 14.7. A student's response to an operations task on the second-semester assessment

THE IMPLEMENTATION PERIOD

To implement the new curriculum for active mathematics learning and its accompanying assessment plan district-wide, the ten teachers who helped develop the plan became campus leaders. As part of the district's focus on site-based management, these teachers presented workshops to engage other teachers in active mathematics teaching and assessment. They shared examples of the results of engaging students in active learning and ongoing assessment, as in figure 14.7. They also shared the feelings they had had and the challenges they had met as they went through the process of change.

As one leader explained to a group of teachers, "It was difficult to change from predominantly paper-and-pencil assessment to hands-on, manipulative activities. At times I wasn't sure I was doing the right thing. Were the children actually mastering the skills and concepts I wanted them to learn? But when I saw the pure joy of discovery, it was worth all the anxiety of change." Others addressed the management problems, with one teacher saying, "One of the challenges I faced was writing down information on a child while questioning and observing him. I also didn't come up with any kind of form to note observations. I've ended up just using notebook paper with each of my student's names on it." Another noted, "Many types of checklists were tried before one was chosen that we felt comfortable with. It took many hours of creating, reviewing, and changing before we found an acceptable plan.... I think there will be more slight changes and revisions as we go along."

Along with presenting workshops to introduce the curriculum and assessment plan, the campus leaders also serve as support personnel for their colleagues as they go through the necessary developmental stages of change, particularly the period of "disequilibrium." As the implementation period continues and this leadership cadre grows to encompass more teachers on each campus, responsibilities for initiating new teachers, program evaluation and revision, and dissemination will be further distributed.

SUMMARY

The goal of this change process was to produce a more appropriate means of assessment for mathematics in grade 1. Teachers are now making use of high-level questions in individual student interviews and their observations of students' behavior, talk, and use of manipulatives within small-group settings in order to assess student learning. This activity-oriented approach to assessment embedded within instruction was developed through a lengthy process that began with instructional changes to generate assessment changes and continued as new assessment requirements produced more changes in instruction. Crucial components to this change process were (1) the access to an accompanying, ongoing instructional project that engaged teachers in a variety of active and integrated mathematics experiences; (2) a small, manageable group of teachers motivated to design an alternative assessment plan and serve as campus leaders; and (3) a mathematics supervisor, or other committed district person, to coordinate the project and facilitate its implementation throughout the district. Although this particular project focused on grade 1 because of its impact on the development of students' conceptual foundations in mathematics, similar projects that provide the necessary components identified in effecting this change in assessment could be successful at any grade level.

REFERENCES

National Council of Teachers of Mathematics. *Curriculum and Evaluation Standards for School Mathematics*. Reston, Va.: The Council, 1989.

_____. *Professional Standards for Teaching Mathematics*. Reston, Va.: The Council, 1991.

15

Just Because They Got It Right, Does It Mean They Know It?

Susan Gay
Margaret Thomas

> *Teacher:* Is 25% of 15 greater than, less than, or equal to 15?
> *Student:* It is less than 15.
> *Teacher:* Would you explain how you decided on that answer?
> *Student:* You subtract. 25% − 15 = 10, and 10 is less than 15.

A S DEMONSTRATED in the dialogue, students may give a correct answer for the wrong reason. If the teacher had not asked the second question, seeking information on the student's reasoning, the teacher would have concluded that the student had an understanding of numbers expressed as percents. This example highlights the purpose of this article, which emphasizes that using only the results of written multiple-choice tests can lead to incorrect conclusions about what a student does or does not know.

Students' understanding of a concept should not be judged by correct multiple-choice answers alone. Open-ended questions and verbal interaction are needed to evaluate a student's understanding of mathematical concepts. This article will illustrate the need for teachers to use these techniques to determine the reasoning a student uses to generate an answer.

In a study with 199 seventh- and eighth-grade students that focused on students' understanding of percent, much was learned about alternative assessment techniques and the value of gathering information from several sources. The assessment techniques included an open-ended question asking each student to explain the reasoning used to answer a multiple-choice question and individual student interviews where the focus was again on the reasoning used by the student. The information gathered from these sources can give the teacher the type of information needed to guide a student's learning. Using these alternative assessment techniques provides detailed, accurate information about each student's knowledge and level of understanding of mathematical concepts.

In addition to the use of alternative assessment techniques, the students in the study completed a written test that contained the following released exercise from the 1986 National Assessment of Educational Progress (Dossey et al. 1988):

Which of the following is true about 87% of 10?

A. It is greater than 10.
B. It is less than 10.
C. It is equal to 10.
D. Can't tell.
E. I don't know.

About 45 percent of the students in the study responded correctly to this question, noting that 87% of 10 is less than 10. This multiple-choice question was followed by an open-ended question that asked each student to explain how he or she decided on a response to the question.

Only about half of the students who correctly chose 87% of 10 to be less than 10 wrote an explanation of an appropriate solution process. An example of the reasoning used by several students is

100% is all of 10 and 87% is smaller than 100%,
so 87% is not all of 10 so it's smaller.

This student correctly compared 87% to 100% and used the fact that 100% of 10 is 10. Another student exhibited a thorough understanding of 50% and 100%, correctly comparing 50% of 10, 87% of 10, and 100% of 10 as shown below.

50% of 10 would be 5.
100% of 10 would be 10.
87% of 10 would be between 5 and 10.
So the answer would be less than 10.

Other students focused on the fact that 87% is less than a whole. One student explained, "87% of 10 is not quite the full amount of 10." Another student used the same reasoning, explaining that because 100% of 10 would be 10, anything less than 100% would be less than 10.

A few students computed 0.87×10 to find the product, 8.7. To support the answer that 87% of 10 is less than 10, one student wrote, "Because 87% of 10 would be 8.7." To justify this conclusion, the student showed "$10 \times .87 = 8.70$." In providing this response, this student demonstrated an understanding of the question, appropriately used a computational procedure, and demonstrated the ability to change 87% to a decimal.

Two students used an estimate of 87% of 10. In response to the question that asked for an explanation for the student's response that 87% of 10 is less than

10, one student wrote, "I looked. 87% of ten is about 9." Another student, using a number line model to estimate 87% of 10, explained, "I thought about how much 87% would be of 10. Almost to the nine on a scale from one to ten. It was less than 10." Through their written explanations, these students demonstrated an understanding of the mathematical concepts involved and an ability to express their mathematical reasoning in writing. As Johnson (1983) indicated, if students can write clearly about mathematical concepts, then they demonstrate that they understand them.

About one-fourth of the students had no written explanation to support their correct choice to the multiple-choice question. This result prompts some concern. It is possible that this lack of response gives some indication of the number of students who simply guessed correctly. It is also possible that these students lacked confidence in their reasoning and chose not to give any explanation. The option to claim that they guessed was an easier way out.

In addition, it is likely that many of the students had not previously encountered questions asking for a written explanation of reasoning in their mathematics classes and were not comfortable writing a response. This contradicts recommendations in the *Standards* that middle school students should have many opportunities to use language in communicating their mathematical ideas (NCTM 1989). Students need to have a reason for making decisions and solving problems in mathematics and the confidence to share that reasoning with others.

Approximately one-fourth of the students who correctly answered the multiple-choice question gave inappropriate explanations. One student using a number-comparison technique wrote

$$10 \text{ is a whole } \#$$
$$87\% = .87$$
$$.87 \text{ is smaller than } 10.$$

This student demonstrated an understanding of some important concepts. The student recognized that 10 is a whole number and 87% is not, that 87% can be represented as the decimal .87, and that .87 is less than 10. However, the student failed to understand the question with regard to the quantity, 87% of 10. Instead, the student compared a percent expressed as a decimal and a number that is not a percent. This reasoning, assuming a correct representation of a percent as a decimal, will work with all similar problems with percents less than 100% and numbers greater than 1. Problems with these numbers are commonly used on multiple-choice tests.

Another student used a counting technique relating the two numbers in the problem, writing, "There's eight as you count up to ten, so I think it's less." This technique may imply a use of division concepts with percent disregarding the percent notation. Even though the technique of counting by the small number to the larger number was used by a few students, it would not have worked if different numbers had been used in the question. For example, if the question had asked about 50% of 5, there would be ten fives in counting by fives to 50 and the response selected would be greater than five. It would surprise teachers to realize how often this counting technique would generate a correct response in a multiple-choice setting, and without a written explanation, its use would not be detected. An awareness of such misunderstanding can help a teacher plan effective strategies for presenting percent concepts.

Other students explained using division, most commonly dividing 87 by 10, though one student did divide 10 by 87. In writing an explanation to support the answer that 87% of 10 is less than 10, one student changed the problem, answering, "Because 10% of 87 is 8.7." To support this conclusion, the student showed 10 divided into 87 and a quotient of 8.7. The student correctly concluded that 8.7 is less than 10. In changing the problem, the student may have ignored the percent sign and focused only on the 87 and the 10, dividing the smaller number into the larger. With this division procedure, these particular numbers generate a quotient that is the same as the product, 87% of 10. Such problems reinforce students' incorrect solution processes, and teachers who check only for correct answers would not detect the lack of correct reasoning.

One student, in describing a general technique that involved the division process, wrote, "You take the numbers and see how many times the smaller number goes into the larger number." The student's reasoning, which incorrectly involves division, had been formed over lengthy work with whole numbers. It is not clear whether the student actually did the division and compared the quotient to 10 or assumed that division gives an answer that is smaller than the divisor. Another student was less specific, noting that "it is less than 10 because of the dividing of the numbers." Division was also involved in the explanation that "all numbers can be divided equally into 100 parts."

It is not clear how a student who wrote "devide [*sic*] half of 87" used this explanation to support the conclusion that 87% of 10 is less than 10. The individual student interview permits the teacher to follow up such student statements by probing for more detail and clarification. Sometimes during an interview the process of talking through thinking strategies seems to correct misunderstandings that otherwise would not be identified from the written responses alone.

Verbal interaction concerning mathematical statements can yield valuable information. When the students responded to the written question "Which of the following is true about 50% of 20?" and were given the choices (A) It is greater than 20; (B) It is less than 20; (C) It is equal to 20; (D) Can't tell; and (E) I don't know, several of them immediately tried to apply an algorithm that they did not understand. However, when the same students were asked during the interview, "What is 50% of 20?" they were able to respond, "10." This would imply that students often perceive written mathematics statements as problems they "need to work out." Without the opportunity to hear a student's verbal response, the teacher would have incorrectly determined that the student could not answer comparison questions using percents. The opportunity for a student to respond in different settings and in different ways is an important aspect of assessment in providing the teacher with the information to facilitate student learning (NCTM 1989).

During the interview sessions, many students who had shown little or no understanding of the percents presented in the multiple-choice questions on the written test were able to explain the meaning of "test scores" of 50% ("one-half of the questions right"), 87% ("almost all the questions right"), and 110% ("all questions right and some extra credit"). These responses support the importance of placing mathematical concepts in appropriate context and using applications of these concepts that are relevant to students. A teacher who had not followed the written test questions with questions of a contextual nature would have concluded that the students had little understanding of percent quantities, when in fact the students only needed help to broaden their practical knowledge to include more general situations and understanding.

Two suggestions for the classroom can help the teacher learn more about students' understanding of mathematical concepts: (1) More one-on-one teacher and student verbal exchanges that focus on the reasoning the student uses to solve a problem can strengthen instruction, and (2) the use of open-ended questions, such as the one in this study, on homework and tests can give students opportunities to practice describing their mathematical reasoning as well as give the teacher additional insight into their level of understanding.

The information gathered in this study of students' understanding of percent supports the need for assessment data from several sources. It is no longer acceptable to rely solely on written instruments that focus on the identification of a single correct response in determining students' understanding of concepts. Written tests provide limited information when presented in a multiple-choice format. Open-ended explanation responses can provide additional information, as long as students are familiar with the questioning technique. Interviews and verbal-response explanations of thinking strategies can supply the teacher with information not obtained through other assessment techniques. Just because they got it right does not always mean they know it.

REFERENCES

Dossey, John A., Ina V. S. Mullis, Mary M. Lindquist, and Donald L. Chambers. *The Mathematics Report Card: Are We Measuring Up?* Princeton, N.J.: Educational Testing Service, 1988.

Johnson, Marvin L. "Writing in Mathematics Classes: A Valuable Tool for Learning." *Mathematics Teacher* 76 (February 1983): 117–19.

National Council of Teachers of Mathematics. *Curriculum and Evaluation Standards for School Mathematics*. Reston, Va.: The Council, 1989.

16

Superitem Tests as a Classroom Assessment Tool

Linda Dager Wilson
Silvia Chavarria

IT IS September. Ann Gomez is planning lessons for a mathematics class of twenty-nine new eighth graders. One of her goals for the class this year is to improve the problem-solving skills of her students. "How much easier and more effective my planning would be," she muses, "if I had a clearer picture of the problem-solving abilities that these students already have. I'd like to have a general profile of the class, as well as a measure of every student's achievements in each of the content areas of the curriculum. At the same time my students shouldn't feel threatened by an extensive battery of tests."

One tool that Ann might find useful is a paper-and-pencil test being developed at the National Center for Research in Mathematical Sciences Education (NCRMSE). (Both authors were members of a project chaired by Thomas Romberg. The purpose of the project was to study the possible use of superitems based on the SOLO taxonomy [see p. 136] for monitoring the growth of students' knowledge of mathematics. Superitems were developed for the content areas of geometry, measurement, patterns and functions, and statistics. Booklets with five superitems each for each content area were pilot tested by approximately twenty-five eighth-grade students [on each booklet] during the fall of 1990 in three middle schools in Wisconsin.) The test is designed to yield class or individual profiles of the problem-solving abilities of students in particular content areas of mathematics (such as geometry, probability, or measurement). Current versions are being developed for grade 8 students, though the idea could be adapted for other levels as well, both younger and older. Each content-area test can be administered to students in a normal

This paper was prepared at the National Center for Research in Mathematical Sciences Education at the Wisconsin Center for Education Research, School of Education, and was supported by the U.S. Department of Education, Office of Educational Research and Improvement (Grant No. R117G00002). The opinions expressed in this publication are those of the authors and do not necessarily reflect the views of the supporting agencies.

class period. The test items are designed to give all students an opportunity to solve at least part of the problem, thus removing the "threatening" aspect that Ann is concerned about. This article will describe such tests, show examples of students' work, discuss how the tests are constructed, and explain how to analyze the results.

The test questions are in the form of superitems. A superitem consists of a situation or story (the "stem") and several questions or items related to it. In the tests designed at NCRMSE, each superitem has four items in each stem. The items represent four out of the five levels of reasoning defined by a theory of developmental reasoning known as the Structure of Observed Learning Outcome (SOLO) taxonomy (for further information on the SOLO taxonomy, see Collis, Romberg, and Jurdak [1986]). All the items can be answered by referring directly to information in the stem and do not depend on the correct response to any previous item. A Level 1 response requires the use of only one piece of information from the stem. Level 2 requires the use of two or more pieces of information from the stem. At Level 3, the student must integrate two or more pieces of information that are indirectly related to the stem, and at Level 4 the student has to be able to define an abstract general principle or hypothesis that is derived from the stem.

According to the theory behind the SOLO taxonomy, it is at the fourth level of reasoning that students are ready for a higher mode of functioning. The development of a student's thinking through the levels constitutes the learning cycle within each particular stage of cognitive development. A pilot testing of superitems that are based on the SOLO taxonomy indicates that, typically, eighth graders are in a transition stage from Level 2 to Level 3, and by grade 12 students are in transition from Level 3 to Level 4. Nevertheless, the example that follows of student responses shows that some eighth graders are able to respond satisfactorily to Level 4 questions.

The following is an example of a superitem that was piloted with grade 8 students:

STEM:

If a figure can be folded so that the two halves lie exactly on top of one another, the folding line is a line of symmetry.

Some figures have more than one folding line of symmetry.

A.

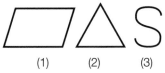

(1)   (2)   (3)

Which of the figures above have folding lines of symmetry?

ANSWER _____

This item requires the use of only one piece of information, which comes directly from the stem (the definition of a folding line of symmetry).

B.

Draw all the folding lines of symmetry on the square.

This item, representing Level 2, requires that the student use the definition of a folding line of symmetry and the fact that some figures have more than one.

C. Which of the first eight capital letters on the alphabet have exactly two folding lines of symmetry?

ANSWER _____

Item C uses the same pieces of information from the stem as Item B, but it requires that the student be able to integrate the information by generating diagrams and applying the definition on a variety of curves. This represents the third level of thinking.

D. John said, "I know a rule for being able to tell when a 4-sided figure has a folding line of symmetry. If the triangles on each side of the line are the same size and shape, it has a folding line of symmetry." Explain why you either agree or disagree with John.

ANSWER _____

_____

_____

To be able to solve Item D, the student has to be able to think critically about a hypothesis that is derived from the stem. This is Level 4 in the SOLO taxonomy.

The format of superitem tests, which are individually worked in class during a specified time period, means that they would not be appropriate for every classroom assessment, since different assessment purposes require different techniques. For example, if Ann wants to determine if students can handle extensive investigations (e.g., How would you wrap a Snickers bar to minimize the amount of paper used?), a paper-and-pencil instrument like the superitem test would be inappropriate. She would gain a much better picture of her students' abilities (and the students would learn much more from the experience) if the problem were treated as a project and the students were actually engaged in finding a solution. But because the superitem format is grounded in a theory of learning and requires only a class period to administer, it can be used for the purposes mentioned above.

Some examples of students' responses to superitems will illustrate the kind of information possible from one item.

For the symmetry superitem described earlier, consider the following two responses from Student 1 and Student 2. Both students answered Items A, B, and C correctly, but they differed sharply in their responses to Item D:

Student 1:   Disagree because not all shapes are made with having triangles in them.

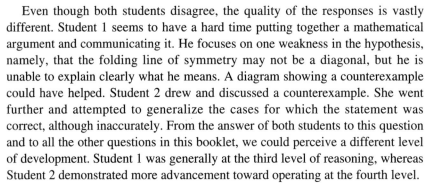

Student 2:   I disagree because, for example, in the rectangle on the left, if you folded on the line of symmetry, the two triangles would definitely not match. Only a square can be folded on the two triangles line of symmetry to be matched up.

Even though both students disagree, the quality of the responses is vastly different. Student 1 seems to have a hard time putting together a mathematical argument and communicating it. He focuses on one weakness in the hypothesis, namely, that the folding line of symmetry may not be a diagonal, but he is unable to explain clearly what he means. A diagram showing a counterexample could have helped. Student 2 drew and discussed a counterexample. She went further and attempted to generalize the cases for which the statement was correct, although inaccurately. From the answer of both students to this question and to all the other questions in this booklet, we could perceive a different level of development. Student 1 was generally at the third level of reasoning, whereas Student 2 demonstrated more advancement toward operating at the fourth level.

Next is a superitem from the content area of measurement, with some examples and discussion of five students' responses.

The measure of the perimeter of a shape is the sum of the lengths of the sides, while the area is the total number of square units contained within the sides.

A. This rectangle has a perimeter of 36 units.

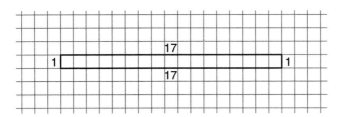

What is the area of this rectangle?

ANSWER _____

B.

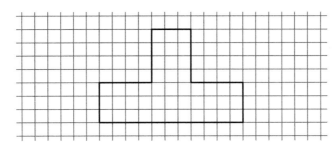

Find the measure of both the area (A) and the perimeter (P) of this figure.

ANSWER   A = _____
                   P = _____

C. Draw an example on the grid below of a rectangle that has a perimeter of 36 units and that has an area larger than 75 square units.

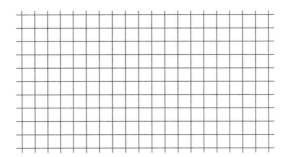

D. On the grid below, there is a picture of the sole of a student's shoe. Most students in a class found an estimated area of this shape by counting the square units. Joe estimated the area of this shoe size in another way. He used a string, which was equal in length to the shoe's perimeter, to form a rectangle. Then he counted the number of squares in the rectangle. Determine if he used a correct method and tell why you think this way.

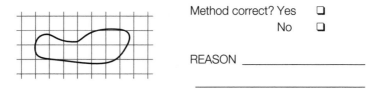

Method correct? Yes   ❏

No   ❏

REASON _____

_____

Students 1 and 2 answered only Items A and B correctly. Student 3 answered Items A, B, and C correctly, but not item D. Two other students (4 and 5) answered all the items correctly. (See the chart in fig. 16.1.)

For Item C, Student 1 drew a 6 × 6 square (a perimeter of 24 and an area of 36), and Student 2 drew a 4 × 9 rectangle (a perimeter of 26 and area of 36). In this item, they both confused area and perimeter, even though both of them were able to compute the area and perimeter correctly in the previous item.

Correct Responses to Perimeter-Area Problem				
Student	A (Uni.)	B (Mult.)	C (Rela.)	D (ExAbst.)
1	X	X		
2	X	X		
3	X	X	X	
4	X	X	X	X
5	X	X	X	X
	Key: Uni. Unistructural			
	Mult. Multistructural			
	Rela. Relational			
	ExAbst. Extended Abstract			
	X Correct Response			

Fig. 16.1. Correct responses to perimeter-area problem

Thus, being able to compute the area and perimeter of a given figure does not imply that a student has a deeper level of understanding of the two concepts, as evidenced by these two students' inability to produce a figure with a given perimeter and area.

Student responses to Item D:

Student 1:   No, because the rectangle would hold 2 more squares.

Student 2:   He used the string to measure his foot; then he could take the string, make the string go around the square units to see how many square units it can go around.

Student 3:   Yes, because if the string was the same length as the shoe perimeter it should work out right.

Student 4:   It is incorrect because if the string was shaped to form a long thin rectangle it would have less area than if it were shaped into a square.

Student 5:   No, because if you lay down the string it will have the same perimeter but it will not have the same area just because it has the same perimeter doesn't mean the areas will be all the same.

The responses of these five students show evidence of different levels of development. Even though Students 1, 2, and 3 answered Item D incorrectly, their responses illustrate the different kinds of misconceptions they hold. Student 1 shows the beginning of some understanding but is unable to fully articulate it, whereas Student 2 simply reworded Joe's incorrect strategy. This student is able to describe a procedure clearly, but does not have a conceptual understanding of the flaws in the argument. Student 3 is willing to accept the incorrect strategy without further explanation. Student 4 gave a correct answer through the use of an example, but Student 5 showed the highest level of abstraction by being able to generalize the solution.

If Ann had these five students in class, what could she learn from studying their responses to this superitem? She now has one piece of evidence about the levels of understanding the students possess in the content area of measurement. Students 1 and 2 are operating at Level 2 on this item. They seem to be able at least to apply the appropriate algorithms for the area and perimeter of a rectangle correctly. Student 3 showed evidence of a Level 3 understanding. He was able to apply his knowledge of area and perimeter to construct a rectangle that met the criteria, but on Item D he showed that he is still confused about area and perimeter. Students 4 and 5 showed some understanding of the two concepts, though there was a qualitative difference in their responses to the final item. Ann should now have some insight about these students' understanding of area and perimeter, and as she studies their responses to all five of the measurement items, she can form an initial profile of their levels of understanding. This should give her data on which to chart these students' growth throughout the school year. It should also give her clues to the kinds of instructional activities she might plan for the year.

In these two examples we have seen how a teacher might use the results from a single superitem to gain valuable information about the level of understanding a student has for some of the concepts of geometry and measurement. A single superitem alone cannot give the teacher a complete picture of a student's conceptions in one content area. However, by studying the results of a student's work on five geometry superitems (or five measurement superitems), a teacher like Ann Gomez would have a fairly clear picture of the strengths and weaknesses of her students' problem-solving abilities in geometry, (or measurement or any of the other content areas of mathematics).

For those who would like to try their hand at constructing their own superitems guided by the SOLO taxonomy, we offer the following considerations, based on our experiences:

1. The construction of a superitem should begin by deciding what general principle will form the focus of the Level 4 item. For example, the general principle in the perimeter-area superitem was establishing the fact that two figures having the same perimeter may have different areas. Once this principle has been determined, the first three items should build toward this principle, which becomes the core of the last item. Each item should help the student explore the problem situation at a deeper level of understanding, forming a unifying whole.

2. The stem should present a problem that is relevant and interesting to students, as well as one that possesses mathematical integrity. No test, no matter how well constructed, should waste students' time with mathematics that is trivial or with problems that are boring.

3. The response to any of the items within a superitem should not depend on the correct response to any previous item.

Our experience in developing and piloting these superitems indicates that if a superitem is well constructed, it should yield fairly reliable results about levels in the SOLO taxonomy. For example, students should get Item A correct if they get Item B correct, and both A and B should be correct if Item C is correct. Results that do not show such an overall pattern probably indicate a flaw in the wording or content of one or more of the items. In the situation of a single student whose responses do not follow the expected pattern, the teacher should be alerted to conduct a personal interview with that student to determine the student's level of reasoning.

REFERENCE

Collis, Kevin F., Thomas A. Romberg, and Murad E. Jurdak. "A Technique for Assessing Mathematical Problem-solving Ability." *Journal for Research in Mathematics Education* 17 (May 1986), 206–21.

17

Facilitating Communication for Assessing Critical Thinking in Problem Solving

Walter Szetela

TEACHERS often hear their students say, "I can do it, but I can't explain it." Doing is important, but students' understanding and communicating what they are doing is more important. If students are able to communicate their thinking, teachers can better assess the quality of the thinking and use the results to help them plan instruction geared closely to their students' needs. This paper addresses the challenging problem of assessing critical thinking.

In Standard 5, the *Professional Standards for Teaching Mathematics* (NCTM 1991, p. 95) makes several recommendations for the evaluation of the teaching of mathematics as problem solving, reasoning, and communicating, including the following:

> Assessment of teaching mathematics as a process involving problem solving, reasoning, and communication should provide evidence that the teacher—
>
> • engages students in tasks that involve problem solving, reasoning, and communication;
>
> • engages students in mathematical discourse that extends their understanding of problem solving and their capacity to reason and communicate mathematically.

These recommendations are related to critical thinking. Effective assessment of critical thinking depends heavily on how well we can facilitate the communication of evidence of students'understanding, critical thinking, and reasoning. In solving problems, aspects of critical thinking include analyzing a problem situation, making decisions, monitoring progress, and evaluating the completed solution. When solving a problem, a student must first obtain an appropriate representation of the problem by considering the problem's facts, conditions, and goal; decide which facts are relevant; and understand how restrictive are the conditions and how clear the goals. When a solution is reached, it must be judged with respect to how well it fits the problem's facts, conditions, and goals. It is difficult to assess such thinking, especially if little has been communicated to assess the quality of thinking.

In one of his practical suggestions for solving problems, Brownell (1942, p. 439) states, "A problem is not truly solved unless the learner understands what he has done and knows why his actions were appropriate." With respect to assessment of the learner's understanding of a problem solution, we could state analogously, "A solution is not truly evaluated unless the teacher understands what the solver has done and knows whether or not the thinking was appropriate."

Thus, to meet the challenge of assessment of critical thinking, we need to furnish problem situations that improve students' abilities to communicate their thinking. We can use well-chosen problems with enhanced formats to promote critical thinking and communication of such thinking, as shown in several examples that follow.

FACILITATING THE ASSESSMENT OF CRITICAL THINKING

The following examples consist of typical problems with supplementary questions designed to encourage communication of critical thinking as a more effective basis for assessment. Examples of students' responses are briefly discussed to indicate the various levels of critical thinking displayed.

1. Withhold the question or a fact from the problem. Have the students examine the problem's facts and conditions and write their own questions and solutions.

Rock music cassette tapes were on sale in a music store. Some were sold for $4 and others for $5. In 10 minutes, 16 tapes were sold.

The teacher forgot to write one more fact. She also forgot to write the question. Make up a useful fact and a question for the problem. Then solve the problem.

A sixth-grade student gave the following responses:

Fact: Altogether $74 worth of tapes were sold.

Question: How many four-dollar and five-dollar tapes were sold?

In assessing these items a teacher will try to determine how well the student is able to assimilate and organize the facts and conditions and how coherently the student's fact and question mesh with the givens. Here the teacher will note that $74 is a quantity consistent with the facts and conditions of the problem. Any amount from $60 to $80 would indicate to the teacher that the student not only had an appropriate representation of the incomplete problem but was able to create an additional fact consistent with the given facts and conditions.

The question created by the student was also consistent with both the given and newly created facts. The relevance and consistency of the two responses furnish more evidence that the student has demonstrated a higher level of thinking and comprehension in the problem than would be shown in the solution to a typical problem completely constructed by the teacher. Because the student was directed to write a fact and a suitable question, the student was encouraged not only to engage in critical thinking but also to communicate information that enabled the teacher to assess the quality of the student's thinking.

Incomplete problem constructions in which students must create facts and questions produce a wide range of responses for teachers to assess. A rudimentary level of comprehension and thinking for the same problem is displayed in another sixth grader's responses:

> *Fact:* How many tapes were in the store?
>
> *Question:* How long was the sale on for?

As well as failing to understand the given facts and conditions, the student did not distinguish the fact from the question. If the student had been asked to solve a completely constructed problem, the extent of the inadequacy of critical thinking may have been less apparent.

2. After students have solved a problem, have them create a similar or related problem.

> We need 6 oranges and 3 lemons to make 8 liters of fruit punch. Your class wants to make 40 liters of fruit punch for Sports Day. Oranges cost 20 cents each, and lemons cost 10 cents each. After you buy the fruit, how much change will you get from $10?
>
> After you solve this problem, create your own recipe. Write a problem using the recipe. Then solve your own problem.

With seventh-grade students, creating problems produces a wide range of responses from simplistic to sophisticated, such as the following (unedited):

> *Simplistic:* 4 bags chocolate chips and 2 bags flour make 12 cookies. A man wants to make 48. how many bags should he buy?
>
> *Sophisticated:* It takes 8 liters of lemon-lime soda and 2 liters of raspberry juice to make 10 liters of raspberry cocktail. Soda costs $1.75 for 2 liters and raspberry cocktail costs $1.15 for one liter. How much will it cost to make 60 liters of cocktail for a large dinner?

In the simplistic construction, the numbers chosen reveal the student's lack of comprehension of real-world quantities and relationships between quantities. Such lack of awareness may reflect a history of artificial school problems in which numerals have been manipulated in computations with little reference to the real world.

3. Present a solution to a problem that contains a conceptual or procedural error or a misrepresentation of the problem. Ask the students to examine the solution and answer a series of questions focused to reveal the extent of their critical-thinking ability. An example of such a problem for fourth grade is shown in figure 17.1.

Donatello, a "teenage mutant ninja turtle," was in the town of Dory at a pizza party with 200 children. His friends, Leonardo, Michelangelo, and Raphael, were in Apex and wanted to join the party. Each decided to take a different route from Apex to Dory. Leonardo took the longest route and Michelangelo took the shortest route. How much farther did Leonardo have to travel than Michelangelo?

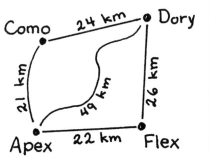

Here is how Jason tried to solve the problem:

$$\begin{array}{ccc} 22 & 21 & \\ 26 & 24 & 48 \\ \hline 48 \text{ km} & \overline{45} \text{ km} & \underline{-45} \\ longest & shortest & 3 \text{ km} \\ Leonardo & Michelangelo & \end{array}$$

Leonardo travelled 3 km more than Michelangelo.

1. Did Jason make good use of all the facts in the diagram? Explain why you think he did or did not.

No, Jason did not make good use of all the facts. He could have used the route that was 49 km for Donatello.

2. If you were the teacher, what would you say to Jason?

I would say Jason do you think you did the right thing? Look closely at the diagram and check your answer.

3. Make up your own question using the facts and information in the problem.

Michangelo went to the pizza parlers in Apex, Flex, Dory and Como. He travelled from Apex to Flex, Flex to Dory, the 49 km route from Dory to Apex, Apex to Como, and Como to Dory. How many km did he drive to go to the pizza parlers?

Fig. 17.1. A problem solution containing a procedural error

Even fourth- and fifth-grade students placed themselves remarkably easily in the role of the teacher, as demonstrated in their responses to question 2. They used language that was more formal, mature, and sensitive, and their responses often imitated those a teacher might make. Following are some unedited examples of fourth-graders' responses:

> Jason this is a good stradgey but you did not look at the route carefully. Go of it again, Jason.

> Jason remember to use your km sign and don't forget the + sign as well. Aside from the answer being wrong it is a good method. Next time try all the information.

A cluster of such questions that focus on a single problem, including questions that place the student in an adult role and that foster creativity, serve as an excellent vehicle for students to engage in critical thinking and to communicate their thinking. The response to the first question in this cluster informs the teacher about the student's ability to analyze and critique a given solution. The second response permits students to demonstrate their selective acuity and judgment when they assume the role of the teacher, and the third response affords opportunities for creativity and demonstration of the ability to coordinate facts, conditions, and relevant questions in a problem the students create.

4. Create a problem for which the student must communicate an explanation without actually solving the problem. The following format, in which fourth graders are asked to explain a problem in a telephone conversation, is based on an idea used in the California Assessment Program (Pandey 1990).

A bus seats 36 persons. At the first stop it picked up 1 person. At the second stop it picked up 2 persons, at the third stop 3 persons, and so on. If nobody gets off the bus, after how many stops will the bus be full?

Suppose that your friend calls you on the telephone and asks you about this problem. What would you say to your friend to help him or her **understand** the problem? **Do not solve the problem.** Just **explain** the problem so your friend will really **understand it**.

Following are examples of students' responses:

> To anser the question you have add up 1 and 2 and so on until you get to 36 and then you count up the stops.

> Well imagine 1 bus ementy. Then 1 person gets on, then 2 people get on, then 3 people get on so on. It's just a pattern count—123456789101112—so on untill the bus is full.

The telephone situation provides a natural setting for the student to verbalize thinking about the problem. When students are induced to verbalize about the problem, they engage in and reveal more thinking than they do when they are simply asked to solve a problem. The written re-creations of the telephone conversations permit the teacher to assess the students' understanding before they solve the problem. The depth of understanding may be revealed by the length of the verbalization, the selection and coordination of facts, the description of a plan for solution, and attempts to relate the problem to materials or experiences more familiar to students. The subsequent solution of the problem and the creation of a related problem elicit additional information about a student's thinking.

SELECTING QUESTIONS TO PROMOTE REPORTS OF CRITICAL THINKING

For purposes of assessment, questions presented *before* students proceed to solve a problem can promote their engaging in and communicating critical thinking. The following are some examples:

1. Do you think this problem will be easy or hard for you? Why do you think so?
2. Do you have difficulty understanding any part of the problem? Describe or explain what you do not understand.
3. Does the problem have any facts or information that aren't needed?
4. Have you solved a problem like this before? Describe such a problem.
5. Can you draw a diagram to illustrate the problem?
6. What strategy do you think will help to solve the problem?

Too often students read a problem and rush blindly into a solution procedure before they understand the problem situation. These questions help students to slow down, think, and obtain better representations of the problem before they choose and implement a strategy.

Supplementary questions posed *after* students solve a problem can also draw out information about students' thinking. Some examples follow:

1. Did you write a complete statement of the answer?
2. Does your answer make sense according to the given facts?
3. What strategy did you use? Why did you choose that strategy?
4. Do you think your solution is correct? Explain why or why not.
5. Was this problem easy or hard for you? Explain why.
6. Could you have solved the problem in another way? Tell how without solving the problem again.

A combination of questions presented before and after students solve a problem can shed more light on students' thinking about a problem, especially one with inconsistencies. An extreme example of such a problem, along with related questions and a fifth-grade student's responses, is shown here.

A Grade 6 class went 18 km by bus to a museum. On the bus were 15 girls and 13 boys. How old is the bus driver?

1. Before you solve this problem, tell what you think about it.

Response: Estimate the answer because this seems to be tough!

2. Now solve the problem if possible.

Response: Older than 16—maybe 43 if you plus all the numbers except 6.

3. Why do you think this is a good or bad problem?

Response: It's good because it keeps you thinking.

In this example the student demonstrates some evidence of critical thinking, such as considering estimation and recognizing that a bus driver must be over 16, but fails to see any incongruity in the problem. Surprisingly, many students in grades 4–7 who observed that the problem was unsolvable nevertheless proceeded to calculate the bus driver's age.

To assess the quality of responses to the questions, teachers can devise brief descriptive criteria for various levels of critical thinking. Although the concept of critical thinking is complex, for practical purposes of assessment the criteria can be relatively simple, as in the following list:

0—The student makes no attempt at critical thinking, indicated by a blank response or a negative comment.

1—The student attempts to answer the question, but the response is illogical or irrelevant.

2—The student understands the question and addresses it with a pertinent comment, but the response is incomplete or confused.

3—The student understands the question and addresses most relevant aspects with correct and logical observations or inferences, or the student addresses all relevant aspects with minor flaws.

4—The student understands the question and addresses all relevant aspects with completely correct logical observations and inferences.

USING SOLVED PROBLEMS FOR ASSESSING CRITICAL THINKING

Typically, students are asked to solve problems. Rarely are they asked to study and critique a given solution to a problem. Their thinking relative to a completed solution may reveal more than the thinking they demonstrate in solving a problem. Pertinent questions constructed to help students communicate their thinking assist teachers in assessing the quality of thinking. Although the problem is already solved, the student must critique the suitability of the strategy and its implementation. If the solution contains a significant error, the error provides a clear and simple basis for assessing students' critiques of the solution.

Figure 17.2 shows an example of a solved tenth-grade problem with a major error and questions that help to reveal the absence or presence of critical thinking, along with an example of excellent responses to the questions (Szetela 1992).

A liter of asphalt paint will cover 6 m² of surface. The paint is sold in cans of 5 L only. How many cans are needed to paint a driveway 15 m long and 3 m wide?

Jill tried to solve the problem this way:

$A = l \times w$

$15 \times 3 = 45$ m² = area of driveway.

$$\begin{array}{r} 7.5 \\ 6\,\overline{)\,45} \\ 42 \\ \hline 30 \end{array}$$

7.5 cans are needed.

Answer these questions about Jill's work:

1. Does Jill's solution show that she understands and uses the problem's facts well? Explain why or why not.

She understands the question about the area but she doesn't seem to understand how to change 7.5 into how many paint cans are needed.

2. Is Jill's answer correct? Explain why or why not.

Everything is correct up to the 7.5. What Jill did not do was seeing that a liter of paint would cover 6 m² and the paint was sold in cans of about 3 L. Only about 2 cans would be needed.

Fig. 17.2. A problem with a major error that helps reveal critical thinking

The tenth-grade student's responses shown in figure 17.2 exemplify the quality of thinking and reporting teachers would like to see. Responses that show little comprehension of the problem and solution and an all-too-willing tendency to accept a given solution without a critical disposition should be no cause for dismay. Inadequate responses may be due largely to students' inexperience with situations requiring critical analysis or to their need for more time to improve the reporting of their thinking. Very important are well-chosen questions that are likely to promote critical thinking and reports of such thinking. Questions suitable for students immediately after they have studied a solved problem might focus on particular aspects of problem solving in which the teacher wishes to assess critical thinking, such as the following:

- *Focus on the reasonableness of the answers.* Does the answer in the problem make sense according to the problem's facts? Explain why or why not.

- *Focus on the selection of the strategy.* Was the strategy used in the problem a good one? Why do you think it was or was not?

- *Focus on alternatives.* Can the problem be solved in another way? Explain how to do so.

- *Focus on the adequacy of the representation.* Did the solver of the problem overlook any problem condition? If so, explain which one was disregarded.

- *Focus on the correctness of the implementation of the strategy.* Did the solver of the problem make any mistakes? If so, explain any mistake.

- *Focus on the goal of the problem, including the units.* Is the statement of the answer complete? Does it contain the proper units?

The difficulty of interpreting and assessing students' reports will be lessened if the questions are very specific because specific questions direct students to address particular targets. Nevertheless, even with carefully selected questions, teachers may be confronted with such vague and uncritical reports as, "Yes, the answer makes sense because that's the way I would do it."

From such a statement it is difficult to determine if the student thought about the answer or was lulled into uncritical agreement with the entire solution. The more unreasonable the answer in the solved problem, the more likely that such a response indicates an inability or unwillingness to examine the given solution critically.

REFERENCES

Brownell, William A. "Problem Solving." In *The Psychology of Learning,* Forty-first Yearbook of the National Society for the Study of Education, Pt. 2, edited by Nelson B. Henry, pp. 415–43. Chicago: University of Chicago Press, 1942.

National Council of Teachers of Mathematics. *Professional Standards for Teaching Mathematics.* Reston, Va.: The Council, 1991.

Pandey, Tej. "Power Items and the Alignment of Curriculum Assessment." In *Assessing Higher Order Thinking in Mathematics,* edited by Gerald Kulm, pp. 39–52. Washington, D.C.: American Association for the Advancement of Science, 1990.

Szetela, Walter. "Open-Ended Problems." In *The 1990 British Columbia Assessment of Mathematics,* edited by David F. Robitaille. Victoria, B.C.: Ministry of Education, Learning Assessment Branch, 1992.

18

Assessment as a Dialogue: A Means of Interacting with Middle School Students

Kris A. Warloe

DECISIONS, decisions. How do people make a decision when they're confronted with having to make a choice? Some people have a very good intuitive sense for making the best choice. However, for the vast majority who don't, skills are needed, and assessments must be made regarding the ability to use these skills to make sound decisions. This means learning to ask the right questions; collecting raw data; organizing, analyzing, and presenting the data in a meaningful way; and making valid interpretations of the data.

The purpose of this article is to present four examples of how I have given my seventh- and eighth-grade students the opportunity to develop these skills, assessed their progress, and assigned a grade. The traditional means of assessing students' progress—pop quizzes, tests, homework, papers—do a pretty good job of sorting and selecting the "top" students who are willing to play the game. However, if we can accept that the primary purpose of assessing student work is teaching and helping the student to learn, then the dialogue that occurs between the teacher and the student, the student and other students, and the student with himself or herself will be beneficial to the student and the teacher as both gain insight into the student's reasoning and thought processes.

STUDENT-GENERATED PROJECTS

As I assess my students' progress, I try to keep two things in mind: What am I saying to the student about what I value, and how will my assessment affect the student? I care about outcomes and want to assess them fairly, consistently, and dependably. It is important that I focus on what the student can do rather than on the number of problems worked or pages done.

The student's ability to use the statistical skills that have been taught can be observed in student-generated projects. I have had my best success when students work on a project in pairs with class time provided. The quality of the students' work depends on well-defined goals, which I call the "Essential Elements" of their projects:

1. A clear statement of the problem to be investigated and a hypothesis or an intuitive feeling for what the results might be

2. The identification of the target population for data collection

3. The method of sample selection from the target population

4. The determination of the sample size; time and cost

5. The survey method; questions, measure, count

6. The method of recording

7. The organization of field work; when and where

8. The organization of data for interpretation

9. Data-analysis techniques

10. Oral presentation

To assess higher-order thinking, assessments must arise from a clear and specific definition of the task. For example, my format for tasks describes the knowledge to be learned and the forms of thinking to be mastered, such as, "Produce various logical interpretations of data sets" or "Debate perceived differences in data sets." The behaviors to be demonstrated also are considered, such as a willingness to persevere on a topic.

The assessment process demonstrates to my students that I value organization and a systematic approach to a project with deadlines to be met. The small increments that are reviewed require a minimal amount of my time, most of which occurs in class while the students are working and I guide their work toward a successful completion of their projects. I maintain a dialogue with my students and conduct short-term assessments as each part of the assignment is completed instead of waiting to "grade" the end product. My students turn in progress reports on where they are in the development of their projects according to the essential elements. I review their progress in short, personal interviews, letting the students know I'm interested and giving direction to their investigation. Requiring an oral presentation to the class sends a strong message that what they do should be their best work.

I use the following general outline for the project and discuss each area with the students as a class.

Due Dates

Day 1: Project directions given; topics chosen

Day 2: First draft of essential elements 1–5

Day 3: First draft of essential elements 6–9

Day 5: Outline for presentations

   Materials needed

   Partner's duties

Day 7: Written report

   Final draft of essential elements, including organized data,
   appropriate displays, inferences from data, problems encountered,
   further possible studies

Days 7–9: Presentations

   Creativity, props

   Delivery—acting out part of the problem

   Information

Students are given feedback as they complete the stages of the project:

• Identify a topic to pursue. Explore something of interest to you.

   Assess how imaginative the students are in choosing a topic. Are they willing to
   take a risk and explore something that is unfamiliar?

• Develop questions to investigate. Determine what is to be investigated.
   Conduct a trial run of your questions to learn if there are unexpected
   responses or other questions that should be asked.

   Assess the students' phrasing of questions. Check for biases in the questions.

• Identify your target population and develop a plan to gather your data.

   Assess how the target population is to be chosen. If it is a supposedly random
   sampling, how is the randomness determined? When, where, and how will the
   data be collected?

• Organize your data.

   Assess how the student records and organizes the data. Are there organized
   charts? Does the student use computer-assisted technology, such as a database
   or a spreadsheet, to help in organizing the data?

• Appropriately display your data.

   Assess the appropriateness of the data displays used. Are the displays accurate?

• Make inferences from your data displays.

   Assess the inferences the students make from their data displays in light of the
   questions asked at the beginning of the study. Does what the students imply
   have any relevance to what was asked? What further investigations are
   suggested by the inferences that are made?

When accepting the challenge of doing a project, students frequently worry
about failure. ("What if our results don't turn out the way the teacher thinks
they should?") When this happens, the students are tempted to alter the data to

produce the supposedly desired results. At the beginning of a project I emphasize to my students that they have not failed if the results of their study suggest something different from their original hypothesis. The ongoing teacher-student dialogue helps to alleviate these fears. What is important is that the students collect the best data possible, analyze the information using appropriate statistical techniques, and make reasonable inferences from the data.

To assess the students' depth of understanding, I look for good graphical displays of their own statistical data. A good graphical display will—

- show the data without distorting what they have to say;
- exhibit the data at more than one level of detail (e.g., from the overall picture to a detailed view of a point of interest);
- encourage the reader to compare different pieces of data;
- be in agreement with the statistical and verbal descriptions of the data set.

CONTEMPORARY PROBLEMS FROM THE COMMUNITY

To assess a student's level of comprehension of dispersion versus central tendency, I try to create a problem relevant to what is currently happening in society, the closer to home the better. We are frequently told by the media the "average" price for a new home. This figure should be questioned. Is it the median, or middle-priced home, or the mean, or average-priced home, in which we are interested? Or does it make a difference? I use the classified section from our community paper to explore this concept. I've found that the "For Sale by Owner" column makes an interesting category to explore, with enough data to make some inferences about our community.

The problem explores how best to communicate to people moving into the community the cost of a home so they can determine the possibility of being able to purchase one. We have learned how to make line plots, stem-and-leaf plots, box plots, and scatterplots. The students use these statistical techniques to analyze the data given in figure 18.1 and develop a brochure for the independent home sellers association that will assist it in encouraging people to move to Corvallis, Oregon, and purchase a home. The project includes photos, floor plans, and price ranges of homes the students think will be desirable to prospective buyers.

This activity lets me assess the student's ability to pick out relevant data (prices of homes that are for sale) from a newspaper; organize the data in a meaningful way; find the mean, median, upper and lower quartiles, extremes, interquartile range, and any outliers; and give a rationale for which data are used in a real-estate promotional advertisement. The student examines the "Homes for Sale" section of the classified ads, looking for advertisements for three-bedroom homes in the local area. The data in figure 18.1 are from the

40,000	46,900	50,000	54,000	57,500	63,000
63,500	68,900	69,500	69,950	72,000	77,900
78,500	82,500	89,950	92,500	95,000	97,500
97,900	98,000	99,500	99,900	118,900	119,000
119,900	126,900	129,900	132,000	136,900	137,000
139,500	141,500	149,900	149,900	151,900	189,500
199,900	205,000	219,000	245,000		

MEAN	112,050
LOWER EXTREME	40,000
LOWER QUARTILE	71,000
MEDIAN	99,000
UPPER QUARTILE	138,500
UPPER EXTREME	245,000
IQR	67,000

Fig. 18.1. Data set on costs of homes and statistics

Corvallis, Oregon, *Gazette Times*, 5 September 1991. The costs of homes are given in dollars.

My instruction is based on an assessment of student understanding. In my dialogue with the student, I am checking for the student's understanding of techniques for statistical analysis. If I determine that the student has a misunderstanding, I can provide feedback that assists the student's learning and leads to a successful product.

Sample Dialogue

T: I see you've found forty homes for sale. Describe for me your general impressions about them.

S: There are a few priced less than $100 000, one that is much more, and several that are more than $100 000.

T: How do you plan to organize your data so that they will be more meaningful to you?

S: A stem-and-leaf plot will help me order the data and find the extremes, quartiles, and median. I can use that information to construct a box plot to show me how spread out my data are. I think I will make a line plot, too. That will show me if there are any clusters of homes in the same price range.

Students want to have the opportunity to make sense of a situation themselves rather than have the instructor give the answer for them. They are looking for guidance on how to interpret the data, but they want to be able to make their own inferences and test their hypotheses. I encourage my students to draw and validate their own conclusions from the data. This process allows my students to think differently about mathematics. Statistical analysis becomes a useful tool to create order in the world. They look for possible reasons for situations that could produce a certain data set and create models that would support their conclusions. I have come to expect a level of discomfort from my students when I ask, "Why? Why do you believe that? Look at the data. What appears to be happening? What inferences can you make?"

EXTENSIONS TO PROBLEMS AND PROJECTS

Projects and contemporary problems almost always lead to spin-off investigations. I pose the following questions to help students become involved in the current affairs of our community and in decisions that may be affecting them in the present or near future. If the median price of a house is $99 000, what would a thirty-year mortgage at the current interest rate look like? If the down payment is 20, 30, or 40 percent of the purchase price, what would the monthly payments be? What are the yearly property taxes? What family income would be necessary if the bank says that it won't loan the money if it takes more than one-third of your income to pay the loan and taxes?

The products created in these types of activities furnish opportunities for student learning. Both the mathematics curriculum and assessment become an integral part of other subject areas, which adds meaning and provides a contextual basis for investigations.

TAKING QUIZZES IN PAIRS

Students' taking quizzes in pairs is a powerful teaching and assessment tool. This assessment technique gives students the opportunity to interact both with me and with each other. At the beginning of the quarter I have each student make a "date card" similar to a dance card for the senior prom. It is in the shape of a clock, with the hours from one to twelve (fig. 18.2). The students are given two minutes to fill out their date card, making a date for each of the hours by getting another student to work with them on that hour. On the day of a quiz, each student takes the quiz independently, recording his or her solution process as well as an answer and submits it for my evaluation. I grade the quiz but make no marks on the student's paper. This is very important so that when the papers are returned, nobody knows whose answers are correct. I maintain a class list, recording which problems are done incorrectly by each student. When I have finished, I have an item analysis of the quiz and know immediately if any areas are causing problems for a large part of the class and need to be retaught.

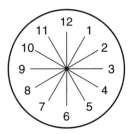

Fig. 18.2

The next day the papers are returned to the students. I put a stem-and-leaf plot and a box plot of the scores on the board for the students to see and interpret. The students are asked to take out their date card and meet for ten minutes with a teacher-chosen date from 1 to 12. When there is an odd number of students or a student is absent, I form on-the-spot groups of two or three.

The ground rules are to sit side by side, use discussion voices, and compare their solutions, including the process. When they disagree on a solution, they share their solution process with each other and try to explain how they arrived at their answer. Neither student knows if he or she has the correct solution, but the interaction between the two students usually develops insights into a strategy for attacking the problem that they may not have had while working independently. A student may then change an answer, provided that permission from the owner has been given, and the student writes down the altered solution process.

After ten minutes the students are requested to meet with a second date at a teacher-designated time, followed by a third date. The students then resubmit their papers to me for another evaluation.

Meeting with three different peers over a twenty-minute time span usually results in most of the errors being corrected. Students who did well initially rarely change a correct answer. They develop self-confidence in their mathematical ability and take pride in explaining how they found a solution. Translating their quiz work to a grade has been relatively easy. The quiz is worth 50 points. It is scored the first time and recorded, and then the process is repeated after they have worked with their partners. The scores are combined and entered as the sum out of 100 points.

I gain information from my students through dialogue between them and me, between themselves, and with themselves. The dialogue is the key to knowing what the students understand. The biggest revelations I've had come from students' journal entries. I've asked students to explain how they would change a percent to a fraction and to give an example expressing their answer as a rational number. More than half the students in class asked me, "What's a rational number?" That was when I decided that frequent dialogue between me and my students using the correct vocabulary was a must.

19

Assessment of Statistical Analysis in Eighth Grade

Maria Mastromatteo

Raw data, graphs, charts, rates, percentages, probabilities, averages, forecasts, and trend lines are an inescapable part of our everyday lives. They affect decisions on health, citizenship, parenthood, employment, financial concerns, sports, and many other matters. (Burrill 1991, p. 3)

THE need for these experiences in statistical training is portrayed throughout current readings in mathematics. The NCTM *Curriculum and Evaluation Standards* places increased attention in grades 5–8 on "using statistical methods to describe, analyze, evaluate, and make decisions" (NCTM 1989, p. 70). Publications such as *Everybody Counts* also stress the need for classroom experiences in this area (National Research Council 1989). To gain facility in using statistics, students must collect, organize, display, and interpret data as part of their classroom experience.

I have taught statistical analysis in my eighth-grade classroom for many years, but within the last three years I have attempted to use procedures that are more in the spirit of the *Standards*. The general goals for which I have worked are to use real data—data significant to the lives of the students; to use good examples to build intuitive skills; and to use cooperative projects whenever possible.

Although this approach is great fun both for the students and for me, it does create a difficult problem—that of assessment and evaluation. Assessment, in my opinion, should do two things. First, and most important, it should measure whether or not the students have met the established goals and objectives. Second, it should provide a means whereby we can justify to parents the grades that have been assigned.

This may sound easy, but in reality it is extremely difficult. The traditional paper-and-pencil test may not measure what we have designed in our lesson. In the past, we gave students a graph and asked them to answer questions concerning the data shown. Although the skill to read graphs is crucial, it is only the beginning. Far more is involved in knowing statistics. Can the students organize, describe, and summarize the data? Can they find patterns? Can they generalize about a larger population or predict an outcome? Following is an example of how the eighth-grade mathematics teachers at our building attempted to teach these skills and the evaluation process we used that proved to be successful for us.

THE "DISASTER WEEK" PROJECT

Our eighth-grade classes decided to do an interdisciplinary unit entitled Disaster Week on four natural disasters. Science, social studies, reading, art, and mathematics were involved. The entire week was devoted to speakers (from the Red Cross, fire department, and state House of Representatives), projects (involving videotape pen pals from a school where there had been a serious flood), films, and lessons. The mathematics classes analyzed data, answering specific questions about one of the four types of disasters that affect our area. We hoped that students could use statistics to make valid decisions, to generalize from a sample of data to the general population, and to communicate information using statistical techniques.

The students were divided into groups of three or four. They were asked questions about the disasters, such as "Which continent is the safest on which to live if you wish to escape an earthquake? Least safe? If we consider only the United States, is it safer to live east or west of the Mississippi River if you wish to escape an earthquake? Does the month of the year have any significance as to when earthquakes happen?" and "Is there a correlation between how many people die in an earthquake and how high it registers on the Richter scale?"

Our assessment purposes were to determine if students could take existing information, use statistical techniques to gain meaning from this information, and then make valid conclusions. Students were permitted to answer any or all of the questions. The students used the data from the *World Almanac, 1990.* The only alteration to the data was that the teachers added the names of the continents beside the country. The students were given specific expectations at the outset. Each group was to produce "a set of well-displayed graphs that answered the questions and demonstrated their knowledge of statistics." The display was to include at least three graphs (fig. 19.1). A written summary that also answered the questions was required. A presentation to the class of each group's findings was the culminating activity (fig. 19.2). Parent volunteers or university students helped those groups that we determined needed extra help with organizational skills.

Teachers assessed students' work on the project using an assessment sheet, which was shared with the students before they began the work (fig. 19.3).

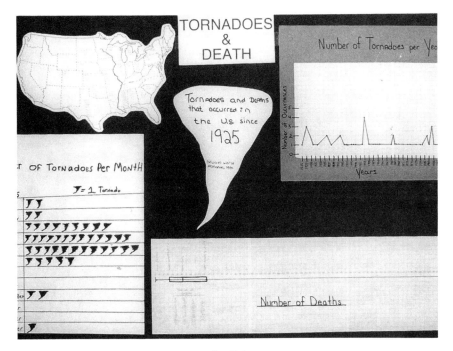

Fig. 19.1

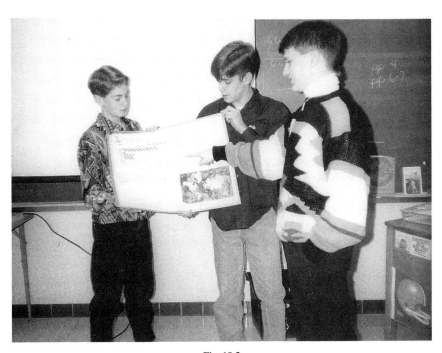

Fig. 19.2

PROJECT EVALUATION FORM

Part I: Graphic presentation

    Graph 1   (10 points)   _____

    Graph 2   (10 points)   _____

    Graph 3   (10 points)   _____

Part II: Conclusion

    Did you answer the questions you set out to answer?

             (10 points)   _____

Part III: Overall appearance

    Visual appeal   (5 points)   _____

Part IV: Presentation

             (5 points)   _____

_____

             TOTAL   _____

             (50 possible)

Fig. 19.3. Project evaluation sheet

In Part I of our assessment form, we checked the validity of the graphs. Were the axes labeled correctly? Were the intervals consistent? Was each graph titled? Was it misleading? Was it unique? Did the graphs show variety—for example, histograms, line graphs, box-and-whisker plots? Within the groups, each student was responsible for at least one graph. The student was required to give an explanation of this graph both in writing and verbally to the class. The student also had to explain what part his or her specific graph played in the overall display of data. These explanations were invaluable in assisting the teachers in determining both the depth of the students' understanding and their ability to communicate their interpretation of the data to the class.

In Part II, the focus was on if their graphs answered the questions that were asked. Could the students translate the information that they had discovered about the entire topic in writing? Not only was content stressed, but also the elements of spelling, grammar, and other writing skills were emphasized. For example, the following brief excerpt is unacceptable: "The month of the year does have a significance as to when floods happen. In some months it rains more than others. This makes rivers, streams, and lakes overflow." This brief excerpt is an example of a conclusion statement in a report. No specifics were cited. The data were either not used or poorly used to make the conclusion. The students should have stated which month had the most floods. They could have used that piece of information plus data about the amount of rainfall during that month to establish a cause-effect relationship—that the rain in those months caused the flooding.

Part III focused on the visual appeal of the display. Was color used judiciously? Was neatness considered? Was there a clarity of concept? An example of a lack of clarity is shown in figure 19.4. This stem-and-leaf chart displays the number of deaths for each flood east or west of the Mississippi River. Each number is labeled E for east or W for west of the Mississippi River. The graph is difficult to interpret. It would have been far clearer if it had been shown on a back-to-back stem-and-leaf plot.

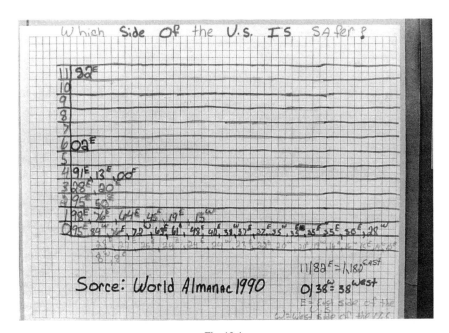

Fig. 19.4

Figure 19.5, however, is very clear in the display of information. Although the graphs are very simple, they clearly show which continent and which month had the most and the least earthquakes. Other questions were also easily answered by simple perusal of the display.

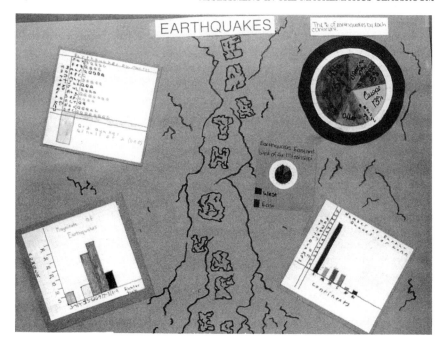

Fig. 19.5

Finally, when the presentations were made, we considered the students' ability to communicate their findings to the class. Once again we looked for a clarity of concept. Was there eye contact with the audience? Was the volume loud enough for everyone to hear? Was there some unique approach to the presentation?

EVALUATION OF THE PROJECT

When the mathematics teachers evaluated the project idea, we tried to see if it met our original criteria. Did it test the students' ability to organize data, draw conclusions, and communicate those conclusions? Each student was asked to fill out an evaluation form for the entire interdisciplinary unit. They rated each activity from 1 to 5. The "Disaster Week" project was one of the activities that the students rated the highest. We believe that this was evident from the amount of time students spent on task and by the lack of discipline problems during this time. We felt that overall the projects did successfully accomplish our objectives. They gave us a measure of the students' understanding, and the presentations showed us their ability to communicate the ideas they discovered through analyzing the data. Our assessment form (fig. 19.3) also gave us a quantitative value that could be changed to a percent that we could communicate to parents.

OTHER SUCCESSFUL APPROACHES

We have found other successful approaches in assessing the learning of statistics. Formal testing that uses data significant to the lives of the students, peer evaluations, and the evaluation of short-term goals have also helped us to determine our students' ability to deal with data and to make informed decisions based on them.

1. *Formal testing* can involve questions that challenge the students' thinking and show their ability to organize and interpret data. For example, Class A and Class B can be compared on which performed better on a specific test. Test scores can be given, and students can find the mean, median, range, mode, and so on. These certainly are data that are significant to their lives. Box-and-whisker plots, stem-and-leaf charts, and line graphs can be made, and written or oral communications can be assessed to show each student's ability to interpret the data.

2. *Peer evaluation* has been a surprisingly valid approach. So long as anonymity is maintained, the students' own evaluation marks, based on a specific scale, are usually quite close to the ones I have given. Students like the idea that this evaluation is part of their grade. They take the responsibility very seriously. We used peer evaluation in two ways.

First, we displayed all the posters (from another class) in the room. The students were to consider the visual appeal and the clarity of the poster's concept. They judged each poster on a scale of 1 to 5 and recorded their assessment on small slips of paper that the groups collated and averaged. The average mark for each poster was used as part of the overall assessment form and contributed 10 percent to their overall grade for the project.

The second way we used peer assessment was through an individual assessment form that each student filled out for every other person in his or her group. Each student was assessed using a scale of 1 to 5 for the following areas:

__Contributed ideas and information
__Acted seriously and maturely about the group work
__Listened to others
__Stayed on task
__Did his or her fair share of the work
__Was cooperative throughout the project
  Comments:

I also evaluated the students using this scale. I collated the results of these assessments and, making sure that anonymity was maintained, shared each student's evaluation with him in a brief conference. Once again I was amazed by the similarity of the students' responses to my own. How they were evaluated by their peers surprised some students and gave them, we hope, some reasons to ponder their input into the project.

3. *Evaluating short-term goals* by allotting points for different parts of an activity also works well. For example, when students choose their own projects, points can be allotted for selecting a question, running a pilot project, gathering the actual data, interpreting the data graphically, and presenting the results. Specific dates can be set up for each part of the project, and a running total of points is kept, with each part contributing to the final total. This total can then be converted to a grade score.

This approach is good for younger students or students who have difficulty grasping the concept of doing an entire project. It gives them a small goal each day rather than the long-range goal of completing the entire project.

CONCLUSION

Although the process of assessment is not easy, it is crucial to the teaching of statistics. The following areas were particularly important to our teachers:

1. Goals and objectives needed to be clear and quantifiable (if possible) at the outset.

2. Directions needed to be clear so that students knew what their specific tasks were. Flexibility in design and presentation should be encouraged but should fall within the parameters set in the directions.

3. The Project Evaluation Form needed to be given to students before they began their project so that they were very familiar with what was expected of them.

4. It was necessary to have a concrete form for recording the results of the evaluation to use to communicate both to students and to parents. The Project Evaluation Form served as this tool.

5. Students' written and verbal conclusions played a most important part in the assessment process.

6. Peer evaluations were important to assess the students' participation and the project's appeal to the students.

REFERENCES

Burrill, Gail, ed. *Guidelines for the Teaching of Statistics K–12 Mathematics Curriculum.* Alexandria, Va.: Center for Statistical Education, 1991.

National Council of Teachers of Mathematics. *Curriculum and Evaluation Standards for School Mathematics.* Reston, Va.: The Council, 1989.

National Research Council. *Everybody Counts A Report to the Nation on the Future of Mathematics Education.* Washington, D.C.: National Academy Press, 1989.

Assessing Reasoning and Proof in High School

Denisse R. Thompson
Sharon L. Senk

REASONING about mathematical concepts and procedures is an important component of mathematical power. Indeed, the importance of this skill is evident in its prominent place in the *Curriculum and Evaluation Standards for School Mathematics* (NCTM 1989). Mathematics as Reasoning is among the first four standards common to each of the three grade-level ranges. Further, the assessment of reasoning abilities is included as an evaluation standard, with an important shift toward emphasizing what students know rather than focusing on what they do not know.

In this article, we discuss four broad issues related to the assessment of reasoning abilities at the high school level. The first is the content about which students are asked to reason. Although high school geometry has long placed a heavy emphasis on proof and reasoning, many other curricular areas can be identified in which justification of conclusions can help foster mathematical power. Thus, we suggest assessing reasoning in algebra, trigonometry, and discrete mathematics, as well as in geometry. The second issue relates to the types of items used to assess reasoning. We illustrate various items and formats, each designed to evaluate some aspect of mathematical reasoning other than just the traditional ability to complete or construct a proof. The third issue we address is how to evaluate students' performance on such items. We present a specific system for scoring open-ended items and discuss issues that arise when such a scoring system is used. The fourth and final issue is the interaction of assessment and instruction. We believe that items and scoring systems similar to those presented in this article lead to insights into students' thought processes

The work reported in this article was partially supported by funding to the University of Chicago School Mathematics Project from the Amoco Foundation, the Carnegie Corporation of New York, and the General Electric Foundation.

and thus provide teachers with critical information on the means to modify lessons so as to facilitate a better understanding of the content.

The illustrative items in this article have been used with students to assess their reasoning abilities as part of evaluations of new curricular materials developed by the University of Chicago School Mathematics Project (Flores n.d.; Hedges and Stodolsky n.d.; Thompson 1992). The responses included here are reproduced exactly as we received them—including spelling and punctuation errors. Names have been changed to protect the identity of the students. Although the items discussed here were used to evaluate the effectiveness of new curricular materials, we believe that they are also appropriate for use in other college-preparatory classes.

All items were scored holistically after modifying procedures described by Malone et al. (1980), Charles, Lester, and O'Daffer (1987); and Senk (1985). Each response was assigned a score from 0 to 4. The meaning of each score is explained in figure 20.1.

Unsuccessful responses

0 Work is meaningless; student makes no progress.

1 Student makes some initial progress but reaches an early impasse.

2 Response is in the proper direction, but student makes major errors; response displays some substance.

Successful responses

3 Student works out a reasonable solution, but minor errors occur in notation or form.

4 Solution is complete; response is fine.

Fig. 20.1. Scoring system for assessment items

Depending on the situation, readers may want to refine the scoring system in figure 20.1. For instance, occasions might arise when teachers would want to distinguish between students who attempt a problem but whose work deserves no credit and students who make no attempt whatsoever on the problem. Or they may want to distinguish between complete solutions that are exemplary and those that are competent. In that case, teachers might find more suitable a scale from 0 to 6 as described in Stenmark (1989).

THE PROCESS OF REASONING

The majority of proof and reasoning items that students experience have such instructions as "Prove the following statement," "Disprove the following," "Show that a is equivalent to b," or "Show that the following expressions are not equivalent." We contend that such directions give students clues to how they should proceed, assuming that students are in tune with the instruction in the classroom. That is, if students know that statements can be disproved by finding a single counterexample, then asking them to "disprove the following" clues them to look for an instance in which the statement does not hold. Likewise, if students understand what is required in writing a proof, then they would not look for particular instances of a statement when told to "prove the following statement is true." In actuality, mathematicians or others who use mathematics do not know in which direction to proceed when faced with a new conjecture. Hence, giving students conjectures to either prove or disprove or items that require them to assess the validity of a given argument is useful. Such items have the potential to give us better insight into the thinking of our students as it relates to reasoning.

Consider the item in figure 20.2 and the responses from students enrolled in a full-year geometry course (Flores n.d.).

Task
A student wrote this "proof" on a test.

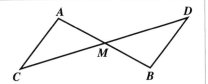

Given: *M* is the midpoint of $\overline{AB}$.
          M is the midpoint of $\overline{CD}$.
Prove: $\overline{AC} \parallel \overline{DB}$.

1. *M* is the midpoint of $\overline{AB}$. *M* is the midpoint of $\overline{CD}$.	1. Given
2. *AM = MB* *CM = MD*	2. Definition of midpoint
3. $\angle A \cong \angle B$	3. All right angles are congruent.
4. $\triangle CAM \cong \triangle DBM$	4. Hypotenuse-leg congruence theorem
5 $\angle C \cong \angle D$	5. Corresponding parts of congruent triangles are congruent.
6. $\overline{AC} \parallel \overline{DB}$.	6. If two lines are cut by a transversal forming congruent alternate interior angles, then the lines are parallel.

A. Which statement best describes your judgment of the student's solution? (Circle your response.)

(i)   It is correct.
(ii) It is not correct.
(iii) I'm not sure whether it's correct or not.

B. Why did you choose the response circled? That is, justify your answer to part A.

Students' responses

I.   (iii) Because the proof is done correctly but the student did not use accurate names for the theorms. (John, score of 0)

II.   (ii) The conclusion of (6) is wrong, it should be "Given". (Sue, score of 1)

III. (i) In the given information, it does *not* say that $\angle A$ & $\angle B$ are right angles, *but* there seems to be enough info for ... $\overline{AC} \parallel \overline{DB}$ (Samantha, score of 2)

IV.   (ii) Because $\angle A$ is not a right angle and neither is $\angle B$. (Carlos, score of 3)

V.   (ii) it states that $\angle A \cong \angle B$ Because Right angles are congruent we do not know if $\angle A$ & $\angle B$ are right angles. (Karen, score of 4)

Fig. 20.2. A geometry item for judging the validity of an argument

To respond to this task, students must critically read the argument and determine whether each justification is valid. This task is quite different from simply proving the theorem on one's own. When an individual proves a theorem, a particular approach or avenue is used. However, evaluating the validity of someone else's proof often requires assessing the appropriateness of an approach not previously considered. Requiring students to explain how or why they determine that a proof is valid or not valid provides different insights into their thought processes.

Consider some of the responses shown in figure 20.2. John's explanation shows that he fails to recognize an important error in reasoning and that he appears to be quite rigid in his use of terminology. We might infer from his response that he believes mathematics can be done in only one correct way. To help students deal with these misconceptions, a teacher might engage students in discussions about equivalent ways to name or state theorems or multiple ways to prove a given theorem.

Sue receives a score of 1. Although she indicates that the argument is not correct, her justification shows that she does not understand the role of given information in a proof.

Samantha catches an important error in the proof, namely that insufficient information is given to conclude that $\angle A$ and $\angle B$ are right angles. However, she indicates that the potential proof is correct and thus confuses the evaluation of the given argument with the ability to prove the conclusion on the basis of the given information. She does not seem to separate her own possible proof from the approach used in the test item.

Both Carlos and Karen have written successful responses. Each states that the argument is not correct and that statement 3 is in error. Karen is correct in saying that "we do not know if $\angle A$ & $\angle B$ are right angles" and thus earns full credit on this item. Carlos mistakenly states that neither $\angle A$ nor $\angle B$ is a right angle and thus receives a score of 3. In general, the difference between scores of 3 and 4 is often a matter of how well students are able to articulate their response.

Figure 20.3 shows sample responses to another item requiring a student to make a judgment. These responses are from juniors and seniors who had studied properties of divisibility (Peressini et al. 1989; Thompson 1992).

Yuri's response displays a fundamental misunderstanding of the process of proof. Although checking the given conjecture for several values to determine its reasonableness is certainly appropriate, such instances cannot prove the statement but can only suggest that the statement is true. Clearly, further instruction is needed to correct this misconception. The teacher may also need to reflect on classroom instruction: How many times do we discuss theorems with students by illustrating the theorem for specific cases? Are such illustrations followed by formal proofs or are the proofs omitted in the name of saving time? Reflection on our instructional practices may suggest how such misconceptions arise and supply valuable information on how to modify such practices.

Task. Is the statement below true or false? If true, prove that it is true. If false, prove that it is false.

For all integers *a, b,* and *c,* if *a* is a factor of *b* and *a* is a factor of *c,* then *a* is a factor of *b · c.*

Students' responses

I. *a* = 4

    b = 8

    c = 12

    4 is a factor of 8

    4 is a factor of 12

    4 is a factor of 8 · 12

    ∴ 4 is a factor of 96.

(Yuri, score of 1)

II. Suppose *a* is a factor of *b* and *a* is a factor of *c*. Then by definition of factors where *m* and *n* are integers, $b = a \cdot m$ $c = a \cdot n$. Therefore, *b · c* can be written as $(a \cdot m) (a \cdot n)$ which equals $a(mn)$ making *a* a factor because *m* and *n* are integers. (Jennifer, score of 3)

III. true. For all integers *a, b,* and *c,* Suppose *a* is a factor of *b* and *a* is a factor of *c*. Then there exist integers *r* and *s* such that $b = a \cdot r$ and $c = a \cdot s$. Therefore $b \cdot c = a \cdot r \cdot a \cdot s = a(ars)$. Since *ars* is an integer, *a* is a factor of *b · c* by the definition of factors. (Lynne, score of 4)

Fig. 20.3. A discrete mathematics item that tests a conjecture

The difference in responses between Jennifer and Lynne lies in attention to detail. Although Jennifer clearly understands the structure and nature of the proof, carelessness results in the omission of the exponent on *a* when she computes $(a \cdot m)(a \cdot n)$. Another minor error in Jennifer's response is her failure to indicate that *mn* is an integer; this condition is necessary in order to use the definition to conclude that $a(mn)$ makes *a* a factor of *b · c*.

Figures 20.4 and 20.5 are examples of open-ended items that address the concept of equivalent expressions or sentences. In figure 20.4, the content is from second-year algebra; in figure 20.5, it is from trigonometry. In each item, the student must decide how to proceed, and both student and teacher should note that no one correct method exists. In the item in figure 20.4, students might expand the first equation, collect like terms, and see if the result is the same as that of the second equation. Or students might begin with the second equation and attempt to put it in vertex form, checking to see whether the result is the same as that of the first equation. In either instance, students must make an initial decision about the necessary process because they do not have initial knowledge about the equivalence of the two equations.

The responses in figure 20.5 also show different valid approaches. Dennis has done a thorough job of finding a counterexample and illustrating how that

---

Task

On a test, one student found an equation for a parabola to be $y - 7 = 3(x + 5)^2$. For the same parabola, a second student found the equation $y = 3x^2 + 30x + 80$. Can both students be right? Explain your answer.

Fig. 20.4. An item from second-year algebra that tests for equivalent equations

---

Task

Prove or disprove the following conjecture.

$$\sin 2x = 2 \sin x$$

Students' responses

I. $x = 90°$

     $\sin 2 \cdot 90 = 2 \sin 90$

     $\sin 180 \ = 2 \cdot 1$

       $0 \neq 2$

    $\therefore \sin 2x \neq 2 \sin x$

(Dennis, score of 4)

II. $\sin (x + x) =$ right side

    $\cos x \sin x + \cos x \sin x =$ right side

    $\cos x (\sin x + \sin x) =$ right side

    $\cos x (2 \sin x) \neq$ right side

    $\therefore \sin 2x \neq 2 \sin x$

(Kathryn, score of 4)

Fig. 20.5. A trigonometric statement to prove or disprove

counterexample disproves the original statement. Kathryn tackled the task by constructing a proof indicating an expression equivalent to $\sin 2x$ and then inferring that $\cos x \neq 1$ for all values of x.

The difference in these two responses points out one of the issues inherent in scoring such items. Teachers must make judgment calls when interpreting responses, particularly when the student chooses a path that the teacher had not considered. For instance, some teachers might want to adjust Dennis's score from 4 to 3 for the use of an equals sign rather than a question mark in the first two lines. Others might be surprised at the approach taken by Kathryn, which is not the typical approach used to disprove a statement, and want to adjust Kathryn's score from 4 to 3 for not making the inference that $\cos x \neq 1$ for all values of x explicit. Still others might want to assign an unsuccessful score to Kathryn's response and consider it an unacceptable disproof of a conjecture. Because the given item was used in an evaluation study in which little knowledge of actual classroom instruction was available, a decision was made to accept Kathryn's approach.

REASONING AND GRAPHING TECHNOLOGY

The *Curriculum and Evaluation Standards* assumes that students in grades 9–12 will have access to appropriate calculators and computers, namely those with the capability of graphing functions. Students with graphing technology can address the question in figure 20.4 in yet another way. They can graph $y = 3(x + 5)^2 + 7$ and $y = 3x^2 + 30x + 80$ on the same viewing window and examine whether the graphs coincide. Figures 20.6 and 20.7 show how the scale on the viewing window of a graphing calculator can influence the perception of whether the two equations represent the same parabola. When graphed on the window with range $-10 \leq x \leq 0$ and $0 \leq y \leq 100$ (fig. 20.6), the graphs may appear to coincide. But when graphed on the window with range $-10 \leq x \leq 0$ and $0 \leq y \leq 10$ (fig. 20.7), the two equations clearly do not represent the same parabola. In general, graphing technology can be used to test whether any proposed identity expressed as an equation in x is reasonable. For instance, to test whether $f(x) = g(x)$ could be an identity, we might graph $y = f(x)$ and $y = g(x)$ on the same coordinate axes. If the two graphs differ at even one point, then the proposed identity is false. If the two graphs appear to coincide, then we can assume only that the proposed conjecture is reasonable; a proof of the identity is then required to verify the conjecture's truth.

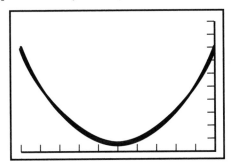

Fig. 20.6. Graphs of $y = 3(x + 5)^2 + 7$ and $y = 3x^2 + 30x + 80$ with range $-10 \leq x \leq 0$ and $0 \leq y \leq 100$

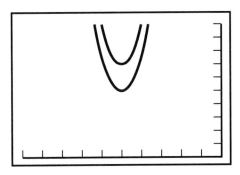

Fig. 20.7. Graphs of $y = 3(x + 5)^2 + 7$ and $y = 3x^2 + 30x + 80$ with range $-10 \leq x \leq 0$ and $0 \leq y \leq 10$

Figures 20.8 and 20.9 contain responses to two items assessing students' understanding of the use of graphing technology in reasoning about identities. The responses in figure 20.8 clearly demonstrate some major differences in understanding. Gladys's response gives little indication that she can even use a grapher with this item. Furthermore, she appears to confuse the terms *identity* and *equation* since her response indicates a desire to find values that make the statement true.

The responses from Sara and Julio, however, indicate real differences in the level of understanding of the required processes. Although Sara recognizes the need to graph each side of the given statement, she indicates no understanding of how to interpret the results. Indeed, she displays a misconception about what it means for a statement to be true.

Julio's response indicates an understanding of the necessary process and the way to determine if the given statement is false. Although successful, he was penalized for notational errors, namely for graphing an expression rather than a function or an equation. This error does not affect the understanding of the process but it requires discussion in the classroom to avoid misconceptions and difficulties later on. Notice that the response of Janice in figure 20.9 contains proper notation by indicating that functions are to be graphed.

Task

Suppose you wonder whether the following statement is true for all values of x.

$$\sin x = \sqrt{1 - \cos^2 x}$$

Describe how you could use an automatic grapher to test the truth of this statement.

Students' responses

I. I would ask somebody for a automatic grapher then I would sit down. then I would open the grapher and adjust the contrast to my angle. Then I would try and figure out how to punch in $\sin x = \sqrt{1 - \cos^2 x}$. I would look at the graph and see if there are any values of x which make this statement true. (Gladys, score of 0)

II. First graph the first statement, $\sin x$ then the second one, $\sqrt{1 - \cos^2 x}$ and if it is a true identity, plug in different numbers for x until you get 2 lines that aren't exactly on top of one another. (Sara, score of 2)

III. First you would use the automatic grapher to draw $\sin x$. Then you can use the automatic grapher to draw $\sqrt{1 - \cos^2 x}$. If the two graphs are different at any point, you know it is not an identity. If the graphs look the same for all points you can assume it is an identity. However, it may or may not still be an identity. (Julio, score of 3)

Fig. 20.8. An item that tests the relation between technology and reasoning about trigonometric functions.

Task

Suppose you wonder whether the following statement is true for all values of $x \neq 0$, 1, or -1.

$$\frac{1}{x} = \frac{1}{x+1} - \frac{1}{x-1}$$

Describe how you could use an automatic grapher to test the truth of this statement.

Students' responses

I. First, you would type in "Graph $y = 1 \div x$" and execute.
    Then, you would type in "Graph $y = (1 \div (x + 1)) - (1 \div (x - 1))$" and execute.

    If the graphs are the same then they are most probably equal. If they are not the same, they are not equal. (Janice, score of 4)

II. graph $y = (1 \div (x + 1)) - (1 \div (x - 1)) - (1 \div x)$ and note that you do *not* get a line at $y = 0$ with holes at 0, 1, and -1. (Olivia, score of 4)

Fig. 20.9. An item that tests the relation between technology and reasoning about rational functions

Another difference can be noted between Julio's response in figure 20.8 and Janice's response in figure 20.9. Julio hints that the graph cannot prove that the statement is true, but his response indicates some uncertainty about this issue. In figure 20.9, Janice clearly expresses the subtlety involved. For the purpose of the evaluation study from which these responses were collected, this subtlety did not have to be noted to receive full credit (Thompson 1992). However, if students indicate that coincidental graphs actually prove that the identity is true, then further clarification in the classroom is required.

Olivia's response in figure 20.9 displays an approach to this problem that had not been considered explicitly in the textbook the students were using (Peressini et al. 1989). This student connected ideas about the intersection of two functions with the ideas relating to identities. Depending on students' backgrounds, the teacher may want to discuss this important connection with the entire class. Once again, the mathematical power of students is enhanced as they view mathematics from many perspectives.

As seen from the students' responses, writing about the process of reasoning yields information that would not be obtained by simply having students determine whether the given statement could be an identity, even if a proof were not required. In a classroom setting, such information provides clues about areas that need further refinement, for example, using equations to indicate the quantities to be graphed or emphasizing that the graph can only suggest, but not prove, an identity. From students who are unable to respond successfully to the item (e.g., Gladys or Sara in figure 20.8), information to guide further instruction can be obtained that would not necessarily be available from simply having students indicate whether the given statement could be an identity.

CONCLUSION

The six items presented in this chapter require students to do more than simply complete a proof or construct a proof "from scratch." They are intended to provide teachers with several models for assessing multiple aspects of mathematical reasoning in high school. In addition, the students' responses to these items generate a wealth of information that can guide the teacher in planning further instruction. Indeed, the richness of the responses generated by such nontypical items cannot be matched by traditional items of the form "prove that...." It is imperative that more nontraditional items be developed if we are to furnish the experiences that help build the mathematical power of our students.

REFERENCES

Charles, Randall, Frank Lester, and Phares O'Daffer. *How to Evaluate Progress in Problem Solving.* Reston, Va.: National Council of Teachers of Mathematics, 1987.

Flores, Penelope. "Evaluation of UCSMP Geometry." University of Chicago School Mathematics Project. Unpublished manuscript.

Hedges, Larry, and Susan Stodolsky. "Formative Evaluation of UCSMP Advanced Algebra." University of Chicago School Mathematics Project. Unpublished manuscript.

Malone, John A., Graham A. Douglas, Barry V. Kissane, and Roland S. Mortlock. "Measuring Problem-solving Ability." In *Problem Solving in School Mathematics,* 1980 Yearbook of the National Council of Teachers of Mathematics, edited by Stephen Krulik and Robert E. Reys, pp. 204–15. Reston, Va.: The Council, 1980.

National Council of Teachers of Mathematics. *Curriculum and Evaluation Standards for School Mathematics.* Reston, Va.: The Council, 1989.

Peressini, Anthony L., Susanna S. Epp, Kathleen A. Hollowell, Susan Brown, Wade Ellis, Jr., John W. McConnell, Jack Sorteberg, Denisse R. Thompson, Dora Aksoy, Geoffrey D. Birkey, Greg McRill, and Zalman Usiskin. *Precalculus and Discrete Mathematics,* field trial ed. Chicago: University of Chicago, 1989. (A revised version of this text is now available through ScottForesman.)

Senk, Sharon L. "How Well Do Students Write Geometry Proofs?" *Mathematics Teacher* 78 (September 1985): 448–56.

Stenmark, Jean Kerr. *Assessment Alternatives in Mathematics.* Berkeley, Calif.: EQUALS, 1989.

Thompson, Denisse R. "An Evaluation of a New Course in Precalculus and Discrete Mathematics." Ph.D. dissertation, University of Chicago, 1992.

21

A Meaningful Grading Scheme for Assessing Extended Tasks

Robert Money
Max Stephens

THIS article describes the assessment and grading of reports of mathematical investigations carried out over an extended time. All eleventh- and twelfth-year mathematics courses in Victoria, Australia, require investigative projects; one such task, based on a centrally set theme, is the first of four statewide assessment tasks normally undertaken by all twelfth-year students of mathematics. One centrally set theme illustrates the teacher's role and shows how a meaningful grading scheme can be applied to assess the product of students' work.

CONTEXT

The Victorian Certificate of Education (VCE) is a set of curriculum and assessment arrangements that apply to all students completing the two postcompulsory years (years 11 and 12) of senior secondary education in Victoria. Full implementation of the VCE in 1992 was the culmination of ten years' discussion and development in which mathematics educators benefited from the wider "total curriculum" context in which their proposals were discussed. This discussion reflected a desire to increase retention rates rapidly and to replace a plethora of examination-dominated and differently valued assessment systems with a single challenging but more broadly accessible set of curriculum and assessment arrangements.

All VCE mathematics courses are developed within a single design for the study of mathematics. This design establishes links among different content areas, work requirements, and a matching set of assessment tasks. A common

The authors acknowledge with appreciation the work of mathematics colleagues at the Victorian Curriculum and Assessment Board, in particular comments and advice provided by Kaye Sentry.

set of assessment standards, covering a wide spectrum of students' achievement, is applied across all twelfth-year courses; four assessment tasks have equal weight in the highly competitive selection for entry into tertiary courses.

CONTENT

The Mathematics Study Design provides a framework within which teachers develop courses, each normally one year in length, that fall into three separate content groupings. One block, entitled "Space and Number," contains arithmetic, geometry, trigonometry, and related algebra. A second block, "Change and Approximation," consists of coordinate geometry, calculus, and related algebra. A third block, "Reasoning and Data," consists of probability, statistics, logic, and related algebra.

Currently, 94 percent of all students include one or two of these blocks in their eleventh-year courses. A growing proportion of students, currently about 75 percent, stay in school for the twelfth year; of these students, about 65 percent take the third block or a block entitled "Extensions," which builds on the content base of eleventh-year work. Approximately half the course content of each block is centrally specified and half can be made up from a range of options.

WORK REQUIREMENTS

The following three work requirements, detailed by the teacher but each taking between 20 percent and 60 percent of the course time, are the basis for determining satisfactory completion of, and certificate credit for, all semester units within Victorian Certificate of Education Mathematics:

1. *Projects:* Extended independent investigations involving the use of mathematics

2. *Problem solving and modeling:* The creative application of mathematical knowledge and skills to solve problems in unfamiliar situations, including real-life situations

3. *Skills practice and standard applications:* The study of aspects of mathematical knowledge through learning and practicing mathematical algorithms, routines, and techniques and using them to find solutions to standard problems

ASSESSMENT TASKS

The three work requirements serve as a framework within which an appropriate range of assessment tasks can be set, either within the school or, for most twelfth-year courses, as Common Assessment Tasks (CATs) administered statewide.

CAT 1, an investigative project, is based on a topic developed by the student within a centrally set theme. The investigative project is undertaken during the first semester of the course.

CAT 2 is a "challenging problem" chosen from four centrally set problems. This activity requires about six to eight hours of work in the problem-solving-and-modeling work requirement and is undertaken over a two-week period in the second semester.

CAT 3, the "facts-and-skills task," and CAT 4, the "analysis task," are ninety-minute tests conducted at the end of the second semester. CAT 3 tests the breadth of coverage of material that should be familiar and routine. CAT 4 is limited to core content but tests higher-order abilities—understanding, interpreting, and communicating mathematical ideas; transferring conceptual understanding to new situations; analyzing complex situations; applying logical argument; and making appropriate checks and estimates.

In Stephens and Money (1991), we have described how these four assessment tasks taken together ensure that an appropriate range of mathematical performance is assessed. This paper is intended to describe in greater detail the administration and scoring of the first task, the investigative project. Many of the comments concerning grading, authentication, and verification also apply to the challenging-problem CAT.

The Investigative-Project Task

Each year a different theme is set for project topics. To assist with course planning, teachers receive information in advance concerning the areas of mathematics involved. Schools receive documentation for the task early in the first semester, and teachers attend a meeting with colleagues from about ten neighboring schools at which their "verification chairperson" introduces them to the grading process and discusses the range of topics that can be investigated under the theme. Students take about twelve weeks to choose their topic and conduct their investigation. The total time spent on the task should be between fifteen and twenty hours (twelve to fifteen hours for investigation and three to five hours for report writing). At least half this time is spent during school hours. Assessment is based on a written report of no more than 1500 words, or ten pages, excluding appendixes.

The Teacher's Role

The teacher assists students in choosing and conducting their investigations, in learning and applying appropriate mathematics, and in communicating their results. Teachers can become involved in—

- organizing access to resources and maintaining liaison with the school librarian;

- planning a three-month timetable that integrates work on the project with relevant skills work and other work requirements;

- explaining the ground rules for the task and the assessment criteria—perhaps through examples from previous projects—with particular emphasis on the criteria relating to mathematical content;

- introducing the theme and possible starting points;

- enabling students to relate these possible starting points to their own interests and mathematical strengths so that they can decide on appropriate topics for their projects;

- helping students to focus their aims for the project and to develop sufficiently detailed and feasible plans of action, including plans for making effective use of time outside the classroom;

- giving advice and feedback to individuals as investigations get under way;

- taking opportunities as they arise to organize group or whole-class learning situations, such as a lesson on mathematics skills that are broadly relevant to the theme and the topics being investigated;

- reviewing the first drafts of project reports and priorities for final efforts;

- supporting students' efforts to turn a set of notes into a written report in the required format. The teacher and student might work through the process together by using a particular example and relating the various sections of the report to the criteria for the award of grades.

Specifications for the report's format include the following:

1. Title page—the exact title of the project, including a summary of aims and findings

2. The main text—a clear statement of the topic and how it relates to the theme, including—

   - an account of any changes of direction after choosing the topic;

   - an account of the conduct of the investigation, including the role of individual students if part of it was conducted by a group;

   - the mathematics and relevant data, including graphs

3. Conclusions—an evaluation of results, discussion of limitations, and ideas for possible further investigation

4. A summary of the mathematical methods used, with examples

5. Acknowledgments of assistance received and resources used

6. References

7. Appendixes—for example, repetitive calculations or tables of data

Project Themes

Together with the statement of the project theme, the documentation includes an introduction to possible starting points for interpreting that theme; see, for example, the following project for reasoning and data (Victorian Curriculum and Assessment Board 1990):

THEME: Choosing the best path

Your project must be based on or incorporate a problem which involves finding or choosing an optimal path. It should use mathematics that is appropriate to the focus of Reasoning and Data. You are encouraged to show initiative and be independent in carrying out your project.

General advice

Choosing a best path is a general mathematical problem that occurs in a wide variety of practical situations. Some best paths really are paths in space, such as the quickest route across a city. Other best paths are not spatial, but are paths only in a mathematical sense. For example, the best sequence in which to arrange contracts to work on a building site can be envisaged as a path through a network which represents how the different building tasks depend on each other and how long each task is likely to take.

As well as clearly demonstrating how the solution to your problem requires you to find a path, you need to make quite clear in what sense the path is "best." Remember that there may be more than one "best" path.

Even if the particular problem you solve may best be solved by trial and error, it is desirable that you also indicate how mathematical techniques could be used to solve more complicated versions of the same problem, which might occur in real commercial situations.

Acknowledge which area of mathematics you use: probability, statistics, logic, algebra. You may choose to develop a computer program to assist you or you may use a recognized computer package but remember to include your own analysis of the problem.

Starting points

You may investigate any topic related to the theme. You must discuss your choice of topic and how it relates to the theme with your teacher. The examples below show some starting points for projects. Note it is not compulsory to use the starting points.

Scheduling of events and examining the likely effects of delays
- Determine a timetable by which to build a house, showing dates for contractors and dates for ordering materials.
- Schedule the tasks in the making of a film or advertisement to be completed by a specific date.
- Timetable the preparation of a meal for a large gathering.

Planning a trip
- Find the shortest or quickest or easiest or cheapest route from one place to a place on the other side of the city by car or by public transport.

- Plan a route around a museum, theme park (e.g., Swan Hill Pioneer Settlement), or a foreign city so that you get to see everything you want to see in a limited time.
- Plan a school trip to the zoo for 5-year-old children (remember to note the feeding times for animals).

Euler and Hamilton paths

- Plan efficient routes for paper delivery, garbage collection, street sweeping or postal service.
- Investigate the history of the travelling salesman problem and its solutions.

Miscellaneous applications

- Design a system to work out daily routes for home delivery of furniture from a large retail store.
- Represent the positions of play in a simple game, such as noughts-and-crosses as vehicles in a tree and investigate and interpret the winning strategies.
- Study the traffic flow from one part of town to another. What difference would a new road make? Where should it be placed to be most cost effective?

Other themes for the investigative project and the year in which they were used include these:

1. Space and number: periodicity (1989), fractals (1990), and circles in design (1991).
2. Change in approximation: paths of moving objects (1989), errors and approximation (1990), and exponential and logarithmic scales (1991)
3. Reasoning and data: predicting uncertain events (1989) and simulation (1991)

Authentication

Student-teacher consultations must be sufficient to enable the teacher to make clear decisions about the authenticity of each student's final report. The "prescribed conditions" give directions that assist teachers in this matter:

1. Students must be given the opportunity to work on their projects during seven to ten class periods distributed throughout the time available.
2. During class periods (and at other times if applicable) teachers will discuss with students their progress on the task and be available for consultation.
3. Students must confirm the following with their teacher:
   - Their chosen topic and its relationship to the theme
   - Their plan for conduct of the investigation
   - The first draft of their report
4. Teachers are required to keep records of the authentication procedures they have implemented.

GRADING AND REPORTING

Project reports, as for all four CATs, are graded and reported on a ten-point scale: A+, A, B+, B, … , E. Two additional "nongrades" are also used. The code UG means "ungraded"—not good enough to meet the requirements, as described in figure 21.1, for a grade of E. The code NA, for "not assessed," may mean that the submission deadline was missed or that the teacher was not able to attest to the authenticity of the report submitted. Grades are awarded according to a set of generalized grade descriptions (see fig. 21.1) that are included on the documentation issued with the certificate at the end of the year.

The development of assessment criteria and grade descriptions was undertaken at the Victorian Curriculum and Assessment Board (VCAB) starting in 1988 with a set of broad policy guidelines and a review of previous local and overseas experience, from New Zealand in particular. Piloting of new courses and all four assessment tasks took place with the cooperation of volunteer schools, initially ten in 1989 and then sixty in 1990. Teachers worked with early drafts of grade descriptions and of the assessment-criteria sheet. The data produced were subjected to computer analysis using "partial credit analysis" software (Masters 1988). This analysis contributed to a review of the grading process and the documentation, which resulted in, for example, the inclusion of the "not shown" category for each criterion. The teachers who participated in 1989 then regraded their students' projects in light of these changes.

The 1990 process led to general confirmation of earlier conclusions by involving students who were completing the second full year of the new mathematics courses and thus were more experienced in project work. The 1990 pilot program confirmed the teachers' ability, within the arrangements developed for the verification process (VCAB 1991), to apply the grading process to project reports originating in schools other than their own. Since 1991 all schools and other institutions offering the certificate have been involved.

Figure 21.1 shows the "Guide to Grade Allocation." It illustrates how the grades and grade descriptions are related in detail to the eighteen assessment criteria that are grouped under three broad headings: "Conducting the Investigation," "Mathematical Content," and "Communication."

Some Grading Exemplars

For the theme "choosing the best path," one student investigated the problem of routes for the four vehicles involved in distributing a newspaper to nine locations from a central depot in a remote country area. The student collected detailed information about required delivery points, time constraints, and the distances involved. The problem was recognized as a version of the "traveling salesman" problem, for which the nearest-neighbor algorithm was studied and then successfully applied. In addition, the student developed and applied the following original "circular route" algorithm:

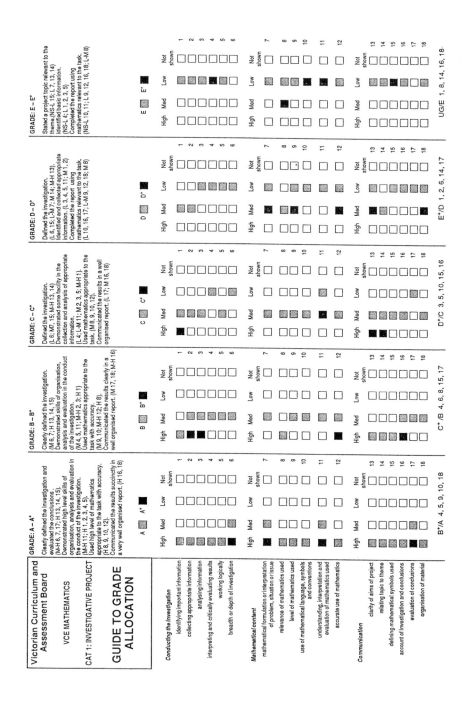

Fig. 21.1. The "Guide to Grade Allocation"

Step 1: Draw the required map to scale.

Step 2: Recognize a circular route around the map.

Step 3: Determine the angles made by the towns not included in the circular route when connected to the route.

Step 4: Incorporate into the circular route the town making the largest angle.

Step 5: Repeat steps 3 and 4 until all towns have been incorporated into the route.

The only two criteria on which the report failed to score a "high" rating—and hence obtained a grade of A rather than of A+—were "breadth or depth of investigation" and "evaluation of conclusions." Key excerpts from the report indicate the deficiencies:

> The nearest neighbor method may not be able to be used where "backtracking" is unavoidable and where it is more logical to use a different route to complete the circuit.

> Of the methods tried, the "nearest neighbor" method was adopted because it didn't require maps drawn to scale, it was simple to perform and it usually resulted in a short, not necessarily the shortest path.

> A question for further investigation might be to work out why the nearest neighbor method does not always give the shortest path—and perhaps find another method which does.

The report contained no discussion of how many paths might be possible in either a specific or generalized situation, only limited discussion (quoted above) of the theoretical status of the nearest-neighbor algorithm, and no recognition of the difference between four one-truck problems and one four-truck problem.

Verifying the Grades

The initial assessment of project reports is undertaken by the school, where the classroom teacher is required to participate in school-level discussion and interaction to ensure internal consistency in interpreting standards and grade criteria. A "VCE Mathematics Assessment Sheet" is completed for each project report, and grade recommendations for each course are finalized by entry into a centrally supplied computer system. This system then makes a random selection of two reports from each grade, and these reports, together with all "A+" and "UG" reports, are forwarded to a verification chairperson for checking against statewide standards and for consistency in interpretation of assessment criteria.

If the chairperson's assessment differs from the school's by more than two grades, all the school's grades at that level might be changed without further sampling. If the chairperson finds inconsistency in the ranking of the sample, then all the school's reports for a particular range of grades may be called in.

Full details of the verification process are documented in the *Verification Manual* (VCAB 1992).

To avoid the foregoing consequences, schools work hard to ensure that their initial recommendations are as accurate as possible, often extending both their curriculum and assessment discussions to local school networks. As a participating verification chairperson in 1992, one of the authors verified approximately 350 assessments emanating from thirty subject groups across sixteen schools. In the end, 95 percent of assessments in these schools were verified without change.

CONCLUSIONS

The assessment technique that has been used with investigative projects demonstrates how a broadly defined set of levels of performance on extended tasks can be consistently applied wherever such tasks are a valued part of the mathematics program—in one classroom, in classrooms in one school, in classrooms in a group of schools, or in classrooms across a state. The grading and verification processes can be used to establish the comparability of grades across schools that is required for public credibility without diminishing the central role of the classroom teacher. Conditions for conducting and monitoring the task can be adapted to meet fairness criteria, including provisions to safeguard the integrity of the authentication process.

REFERENCES

Masters, Geoffrey N. "The Analysis of Partial Credit Scoring." *Applied Measurement in Education* 1 (4) (1988).

Stephens, W. Max, and Robert Money. "The Range of Performance Assessed." Paper presented at the ICMI Study on Assessment in Mathematics Education and Its Effects, Calonge, Spain, April 1991.

Victorian Curriculum and Assessment Board (VCAB). *Investigative Project CAT 1: 1990 Reasoning and Data.* Melbourne: VCAB, 1990.

_____. *Mathematics Course Development Support Material.* Melbourne: VCAB, 1990.

_____. *Mathematics Study Design.* Melbourne: VCAB, 1990.

_____. *Student Material to Accompany the 1991 VCE Mathematics Verification Manual.* Melbourne: VCAB, 1991.

_____. *Verification Manual.* Melbourne: VCAB, 1992.

22

An Authentic Assessment of Students' Statistical Knowledge

Joan B. Garfield

PROBABILITY and statistics are increasingly being given an important place in the K–12 mathematics curriculum. According to the NCTM *Curriculum and Evaluation Standards* (1989), students should learn to apply probability and statistics concepts to solve problems and evaluate information in the world around them. The statistics standards suggest using hands-on activities involving collecting and organizing data, representing and modeling data including the use of technology, and communicating ideas verbally and in written reports. Teachers are encouraged to help students develop important ideas (for example, about distributions, randomness, and bias) and gain experience in choosing appropriate techniques to use in analyzing data.

Many teachers are currently using materials from recent projects or projects in development that have developed curricula and software to implement the NCTM Standards (e.g., the Quantitative Literacy Project, the Reasoning under Uncertainty Project, and the ChancePlus Project). These new materials encourage teachers to have students work on statistical projects: formulate research questions, collect and analyze data, and write up the results. Working on statistical projects individually or in groups engages students in learning about statistics and helps them to integrate the knowledge they have learned.

The Need for New Assessment Methods for Statistics

Historically, most mathematics testing has focused on students' computational skills and few tests have measured higher-order thinking. Statistics items that appear in traditional tests typically test students' ability to calculate correctly the mean and median for a set of numbers or to read a number from a graph. Tests composed of these items not only test skills in isolation of a problem context but do not test whether or not students understand how these statistical measures are interpreted or know when one is a better summary measure to use than another. They also fail to assess students' ability to integrate statistical knowledge to solve a problem and their ability to communicate using the statistical language. Alternative forms of assessing statistical learning are needed that will inform teachers about how well students can communicate using the statistical language, understand statistics as an interrelated set of ideas, and how well students are able to interpret a particular set of data.

In reviewing the NCTM Standards for assessment, Webb and Romberg (1988) provide criteria for new assessment instruments for mathematics. These criteria can also be applied to the development of statistical assessment materials. Such instruments should—

1. provide information that will help teachers make decisions for the improvement of instruction;

2. be aligned with instructional goals;

3. provide information on what students know;

4. supplement other assessment results to provide a global description of what mathematics students know.

The "Practical Project" as an Assessment Method

The "practical project" was originally designed as a learning activity to help students integrate what they had learned in preparing for a cumulative exam. In grading these problems, I learned that they are extremely useful indicators of students' understanding of statistical ideas and their ability to apply these ideas in analyzing data and that they offer valuable insights to the teacher about where additional instruction is needed.

There are two versions of the practical project. In the first version, students collect a set of data of interest to them (consisting of 20 to 40 values) that they will be describing and exploring. Many ideas for data sets are presented and discussed in class, and students are strongly encouraged to gather data in which they are really interested. Examples are given of different types of sports data (data for various teams or individual team members) or data related to popular music (number of minutes of songs, CD prices at different stores). Students are encouraged to look for data sets in magazines (e.g., *Car and Driver, Consumer Reports*), almanacs, and the newspaper. Some students choose to collect their own data rather than use an existing data set. One student decided to find out the cost of a dozen roses at a variety of florist shops in the city. Other students have collected data at part-time jobs, using sales receipts, tips earned waitressing, or gasoline sales. This is a good opportunity to exploit student diversity.

In the other version of this project, students collect information about themselves every day for three to five weeks. Types of data collected have been the number of minutes spent each day on homework, talking on the telephone, or watching television or the amount of money spent or earned each day. Before these projects are begun, it is important to have each student submit an explanation of what data she or he has selected or decided to collect along with a sample of the data values. The teacher should examine the data values to make sure they are quantitative data and have enough variability to be worth analyzing. Students are instructed to take accurate measurements of time, money, and weight, and not round off their data values to the nearest whole number.

Guidelines for Data Analysis

The analysis of the data can be done individually or in groups (the group selects a particular data set to work on together). Students should use calculators for calculating statistical measures. They should use the following guidelines both in analyzing the data and in reporting the results:

1. Describe the data you collected. What do the data represent? Why did you choose these data. How did you collect them?

2. Summarize the data in a table and use different types of plots to graph the data.

3. Calculate appropriate summary statistics for your data including measures of center and variability. Show your calculations.

4. Write a description of your data set.

   a. Describe the information revealed by the plots, the shapes of the distributions, and how different plots give different information about your data set.

   b. Interpret the summary statistics in terms of the data. Specifically, explain what the measures of center indicate and how the different measures compare.

   What information do the measures of variability give?

   c. Does your distribution look like a normal distribution? Why or why not?

5. What did you learn about your data? What questions do you have about your data now that you have analyzed it? What would you do differently next time? What other variables that relate to this one might you choose to analyze?

Evaluating the Practical Project

An analytic scoring method, adapted from the holistic scoring method offered by Charles, Lester, and O'Daffer (1987) for evaluating student solutions to mathematical problems, can be used to grade the projects. A score of 0 to 3 points is assigned in each of the following categories:

- Communication
- Visual representations
- Statistical calculations
- Decision making
- Interpretation of results
- Drawing conclusions

According to this rating scheme, 3 points indicates correct use, 2 points indicates partially correct use, 1 point indicates incorrect use, and 0 means there was not enough information to evaluate or that this part of the project was missing or incomplete. Because this is a project that requires a large amount of student effort both in analyzing data and in writing up the results, this modified scheme still allows students to earn 1 point for each component if the effort was made but was incorrect. Explanations of the categories and scores along with samples of students' work are given below.

- *Communication:* The appropriate use of statistical language and symbols

Score	Interpretation
1	Inappropriate use: student misuses words or symbols or uses statistical words in a context that does not make sense.

Example: *"This data that I collected tells me that the central value measurements were the best way to formulate these data. Most of the information given was very close to the average and was a better way to represent this material."* It is hard to figure out what the student is trying to communicate here. Several different ideas seem to be confused.

2	Partly appropriate use of language or symbols

Example: *"The average and the trimmed average are quite comparable. I feel the average works best for me. The median is slightly higher, it is the center of all my times but on most nights I got less than 7 hours of sleep; not more.... The center of the data can be different depending on how you find it."* There is some confusion in describing the measures of center and using them in interpretive statements.

3	Correct use of language and symbols

Example: *"The histogram is skewed to the right or to the higher values. The histogram and the box plot give me the same type of information. There are no outliers and the majority of the values are at 20 or below. The box plot does give me more information, showing that 50% of the values are below 10 and 25% are between 10 and 20."*

• *Visual representations:* The appropriate construction and display of tables and graphs

Score	Interpretation
1	Errors in constructing or labeling tables and graphs. For example, leaving out a vertical scale for a histogram or failing to use an equal interval scale for a box plot as shown below:

Note: Although the medians and quartiles might be correctly calculated for the box plot, the score of 1 reflects errors only in graphs. Calculations receive a separate score. Other types of errors might be using bars with unequal widths in histograms or having sequential bars not touching each other.

2	Tables and graphs have some errors but are mostly correct.

Figure 22.1 shows the introduction, stem-and-leaf plot, histogram, and box plot for Student A. This sample indicates that the student is somewhat weak in constructing graphs. Although the stem-and-leaf plot is correct, it is missing a code and label that tell how to interpret the numbers and what they represent. The traditional histogram is lacking a vertical scale and the horizontal scale is not correctly labeled. The box plot appears to be correct, although it is sloppily drawn. I would give this student a score of 2 for visual representation.

3	Tables and graphs are correctly constructed.

Figure 22.2 shows a sample of another student's work. All graphs are correctly constructed and labeled. I would give this student a score of 3 for visual representation.

• *Statistical calculations:* Statistical measures are calculated correctly and appear reasonable.

Score	Interpretation
1	Errors in calculation lead to answers that are unreasonable, or a formula is not used correctly. For example, students use the median rank instead of the actual median data value, fail to average the middle values in finding the median, give a negative value for a variance or standard deviation, or calculate a mean that is obviously larger or smaller than the rest of the data values.
2	Some correct, some incorrect calculations
3	Calculations appear to be done correctly.

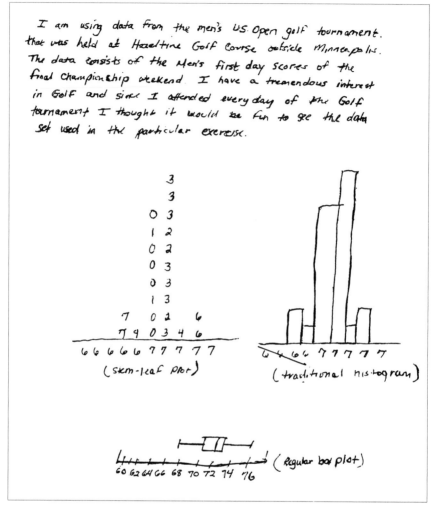

Fig. 22.1. Portions of Student A's practical project

• *Decision making:* The tables or graphs selected are appropriate for representing the data; the appropriate summary measures are calculated.

Score	Interpretation
1	Inappropriate tables or graphs are constructed, and decisions do not appear to be based on obvious clues in the data. For example: a histogram is made with 20 short bars indicating a bumpy pattern and no apparent shape instead of condensing it to a fewer number of bars to better reveal the shape of the distribution.
2	Some decisions seem inappropriate, whereas others are appropriate.
3	All decisions appear to have been made correctly.

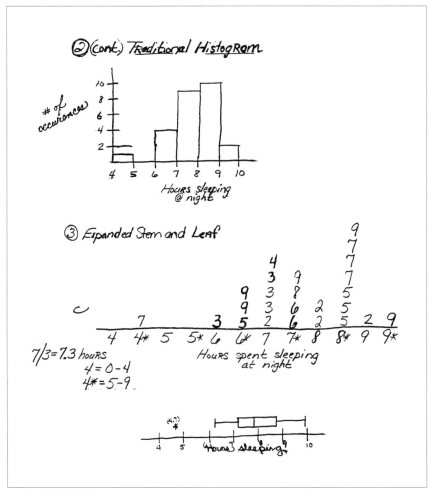

Fig. 22.2. Portions of Student B's practical project

• *Interpretation of results:* The ability to use information from representations and summary measures to describe a data set

Score	Interpretation
1	The student seems unable to interpret the plots and measures. For example, the student fails to recognize the shape of the distribution (when it is obviously skewed, bell-shaped, rectangular, etc.), makes unreasonable or incorrect statements (e.g., "The standard deviation and the interquartile range are too close to the average" or "The average is a little smaller than the median, which would mean there are more smaller numbers than $50 than numbers larger than $50."), describes a distribution as normal when the analysis indicates it is

not normal, or merely restates information without interpreting what it means.

2   Interpretation is too brief, the student fails to interpret some important information, or the interpretation is partly correct and partly incorrect.

3   Good interpretation of data set using all appropriate information

Here are examples of good interpretations from students' papers:

"From the shape of the histogram and the fact that the mean, median, and mode are all close together, the data seem to have a normal distribution."

"An outlier with a large value influenced the mean."

"The trimmed average and the average are very close, meaning there are no outliers."

"My data set consisted of how much money I spent each day for 29 days. My graphs show that my data values are skewed to the right, which means I spent a lot of money just a couple of times while I spent a little money many times. The mean says I spent about $14 a day, but I didn't. I spent maybe no money one day while I spent $50 the next day. The median seems more reasonable, saying that I spent less than $4.50 a day for half the days and more than $4.50 a day for half the days."

• *Drawing conclusions:* The ability to draw conclusions about the data, point out missing information, or relate this study to other information

Score	Interpretation
1	Student fails to draw conclusions or draws conclusions not substantiated by the data. Or the student fails to notice inconsistencies in conclusions made.

Example:

A student who studied the number of minutes slept each night reported a range of 375 minutes and a variance of .05 minutes. The student concluded, "Although the range was quite large, the variance and standard deviation prove that I get a very similar amount of sleep (.05 variation) over a three-week period." It is clear that the student did not see the contradiction implied by a range of 375 minutes and a variance of .05 minutes. The student also erroneously believes that the measures "prove" something: that he got a similar amount of sleep each night.

2   Weak conclusions are made, but some attempt is made to look beyond the data set.

3   Conclusions are made on the basis of the data analysis. Comments are made about the study as it relates to the real world.

Example:

"My data set suggest that I usually get either 3–4 hours of sleep or 7–8 hours of sleep. These are the two peaks in my data. I find this interesting and it also makes sense to me—since I work every weekend morning, I rarely get sleep on the weekends. I try to catch up more during the week."

Scoring Students' Projects

I have designed a rating sheet that I use to score student projects. This sheet is stapled to each student's paper and provides information both on the scoring categories and on why students lost points. Figure 22.3 shows a student's rating sheet.

Practical Project Rating Sheet   Name _Sarah H._____

1. Appropriate use of statistical language and symbols: (3 pts.)_3__
   Comments:

2. Appropriate construction and display of tables and graphs: (3 pts.)_2____
   Comments: _Missing code and label for stem and leaf plots._

3. Correctness of statistical calculations: (3 pts.)_3_____
   Comments:

4. Appropriate choice of tables, graphs, and summary statistics: (3 pts.)_2____
   Comments: _Should have used a regular boxplot to show outliers_

5. Reasonable descriptions and interpretations of data: (3 pts.)_3__
   Comments:

6. Appropriateness of conclusions: (3 pts.)_2___
   Comments: _You did not discuss the outliers and how they affected your analysis and results._

Total score: (18 pts.)_15_

Fig. 22.3. Sarah's rating sheet

How the "Practical Project" Meets the Criteria for New Assessment Methods

The project provides information on what students know. Scores may be given to students for each category as well as a final score ranging from 0 to 18 points. The individual scores indicate to students their strengths and weaknesses, so that they can improve weak areas. Teachers wishing to assign a letter grade on this problem would be cautioned not to base this grade exclusively on the percentage of total points earned.

It provides information that will help improve instruction. Teachers can use the category scores to judge where more activities and instruction are needed.

It can be aligned with instructional goals. The "practical project" as an assessment method is aligned with the instructional goals stated in the NCTM *Standards.* Students are engaged in the basic activities of data exploration by conducting investigations involving higher-order thinking and communication.

The project helps describe what mathematics students know. This method provides useful information about how students use their statistical knowledge and skills in solving an applied problem. The results can supplement other assessment information (such as quizzes and tests) to furnish a better understanding of what statistics students know, and the results can be combined with other assessment information to provide a broader description of the mathematics students know.

An added benefit is that students often learn something interesting about the world or about themselves as a result of working through the project. This is especially true if they have analyzed data about themselves. Students have commented on discoveries made about their habits of spending money, their sleep patterns, and their television watching. Students who have collected data on car performance or flower prices have commented that they learned something interesting they did not know before about the variability of those data. Another benefit is that although students initially may express reactions about the problem's being too long, too hard, or too much work, they often comment afterward that the process was very helpful in preparing for a test, in figuring out what to study, and in pulling together what they had learned. Students also show a sense of pride in being able actually to "do" statistics.

REFERENCES

Charles, Randall, Frank Lester, and Phares O'Daffer. *How to Evaluate Progress in Problem Solving.* Reston, Va.: National Council of Teachers of Mathematics, 1987.

National Council of Teachers of Mathematics. *Curriculum and Evaluation Standards for School Mathematics.* Reston, Va.: The Council, 1989.

Webb, Norman, and Thomas Romberg. "Implications of the NCTM Standards for Mathematics Assessment." Paper presented at the AERA Annual Meeting in New Orleans, 1988.

23

Assessment in Problem-oriented Curricula

Jan de Lange

MATHEMATICS education is changing fast in a number of countries. In certain countries—the Netherlands, Denmark, and Australia, to name a few—these changes have already taken place in the eighties. Other countries are changing their mathematics education in the early nineties.

CHANGING MATHEMATICS EDUCATION

Changing Goals

The changing conditions have led to changing goals. In the Netherlands, for instance, the goals (for the majority of the children) are—

1. to become an intelligent citizen (mathematical literacy);

2. to prepare for the workplace and for future education;

3. to understand mathematics as a discipline.

These goals very much resemble the set of goals stated by the British Committee of Inquiry into the Teaching of Mathematics in Schools (Cockroft 1982). The goals reflect a shift away from the traditional practice. Traditional skills are subsumed under more general goals for problem solving, communication, and critical attitude.

Changing Theories

At the same time the goals of mathematics education are changing, we also see new theories evolving for the learning and teaching of mathematics. At the Freudenthal Institute (formerly Instituut Ontwikkeling Wiskunde Onderwijs

[IOWO] and Onderzoek Wiskundeonderwys an Onderwijs Computer Centrum [OW&OC]) the theory of "realistic mathematics education" evolved after twenty years of developmental research that seems to be related to the constructivist approach (see Freudenthal [1983, 1991]; Treffers [1987]; Lange [1987]; Gravemeijer, van den Heuvel, and Streefland [1990]; and Streefland [1991]). Some differences, however, can be cited.

The social constructivist theory is in the first place a theory of learning in general, whereas the realistic mathematics theory is a theory of learning and instruction in mathematics only. One of the key components of realistic mathematics education is that students reconstruct or reinvent mathematical ideas and concepts by exposure to a large and varied number of real-world problems and situations that have a real-world or model character. This process takes place by means of progressive schematization and horizontal and vertical mathematization. The students are given opportunities to choose their own pace and route in the concept-building process. At some moment, abstraction, formalization, and generalization take place, although not necessarily for all students.

Changing Content

Not only goals and teaching and learning theories have changed mathematics education. New subjects, most notably discrete mathematics, are slowly and sometimes reluctantly introduced into curricula, and interest in geometry seems to have been revived.

Some of these subjects enter the curriculum because new technology has opened new possibilities. The computer has had some limited impact on the teaching of mathematics, but future development might have more visible effects. A graphing calculator with a computer algebra system would outdate both personal computers and graphing calculators as we know them.

Changing Assessment

A lot of truth can be found in the conclusion of Galbraith (1992) that constructivism drives curriculum design and the construction of knowledge but positivistic remnants of the conventional paradigm drive the assessment process. In the Netherlands we boldly confronted this contradiction. Many teachers and researchers react, "I like the way you have embedded your mathematics education in a rich context, but I will wait for the national standardized test to see if it has been succesful."

During experiments that eventually led to new curricula in the Netherlands for the upper-secondary level, we were confronted with two serious problems: (1) Time restricted written tests (these are open-response questions and not multiple choice) were improper for testing, especially for testing the higher-level goals. (2) Under any conditions, designing proper tests is very difficult. When designing tests was left to the teachers, the results were disappointing.

Only 15 percent of the exercises really tested other than the lowest level. Consequently we started developmental research to test new formats. We adopted these guiding principles:

1. Tests should be an integral part of the learning process, so tests should improve learning.
2. Tests should enable students to show what they know rather than what they do not know. We call this "positive testing."
3. Tests should measure all goals.
4. The quality of the test is not dictated by its possibilities for objective scoring.
5. Tests should be practical enough to fit into current school practice.

These criteria were later also used in the test development of the MORE project for the elementary school level, as van den Heuvel points out in her contribution in this book.

NATIONAL EXAMINATIONS IN THE NETHERLANDS

In the Netherlands, final nationwide examinations are administered at the end of four, five, or six years of secondary education. Roughly, one can say that the six-year curriculum prepares the student for the university, the five-year curriculum for higher vocational training, and the four-year course for the lower vocational level. Otherwise students can start working right after their secondary education.

Since 1985 we have had two new curricula for the six-year course. The experiments (carried out by OW&OC) that led to the new curricula showed very clearly the problems of achievement testing. The first nationwide examination on the new curricula took place in May 1987. Before 1987, examinations were time-restricted written tests (TRWT) with so-called "open questions" and no multiple choice. However, the open questions were in fact very closed because both the answer (a number, a graph) and the solution left no freedom of choice whatsoever. Since 1987, modest progress has resulted in a somewhat more open final examination, still of the TRWT kind, but more problem and text oriented.

In the summer of 1990 two new curricula for the five-year stream were introduced: one (B) for the more mathematically gifted students and another (A) for those who are going to use their mathematical skills in their nonmathematical profession or schooling. The first experimental examination on these curricula took place in May 1989. The results of the examination showed that indeed some progress was being made. The exercises—

1. are more open to multiple solution strategies and results;
2. assess higher-order thinking skills;
3. give the students the freedom to solve problems at their own ability levels.

New Curricula and Assessment for Ages Twelve to Sixteen

In August 1993, new mathematics curricula for students aged twelve to sixteen will be introduced if no political disaster happens. Again some five years of experiments have led to these new curricula. Major changes have been accomplished: An "adjustment" was made in the algebra part of the curriculum, stressing reasoning and mental processing and de-emphasizing tricks, manipulations, and drill. For geometry the change is even more drastic: away from the abstract, partly deductive system of discussing triangles and rectangles and toward the understanding of three-dimensional space and the geometry of shadows, perspective, maps, shapes, orientation, and much more. Integration within mathematics is the focal point; talking about algebra and geometry as isolated subject matter is definitely not a proper way to look at the present developments.

Assessment was emphasized as part of the experiments. Some relatively small experiments took place on the national level with the Central School Examination. It may be interesting to look at some of the exercises in, and the results from, the 1992 examination.

We have first chosen an example that makes clear that, contrary to the belief of some people, the Netherlands examinations do include problems with no context (bald or bare problems), or to put it differently, mathematics is the context. It shows to a certain extent the changes in algebra, in this instance, for the lower track of students of about sixteen years.

Examples from the Examination for Twelve-to-Sixteen-Year-Olds

The table (see worksheet [table 23.1]) gives values for $x^3 + x$ and $1000/x$.

1. Take for x the values 2, 5, and 8 and complete the table.

2. Compute the difference between $x^3 + x$ and $1000/x$ for $x = 100$.

3. Draw (or make a sketch of) the graphs of the functions $x^3 + x$ and $1000/x$ using the numbers from the table.

4. Find a value for x for which $x^3 + x$ and $1000/x$ are about equal in value.

5. Explain how to find a value for x for which the difference between $x^3 + x$ and $1000/x$ is less than the difference in question 4.

TABLE 23.1

x	$x^3 + x$	$\dfrac{1000}{x}$
1	2	1000
2		
5		
8		
10	1010	100

The participating teachers and the developers were surprised that this item was done very well by the students because of the perceived lower ability level of these lower-tracked students and also because the problem was not presented in a "nice" context. The desired "sketch" quite often really was just a sketch, as illustrated in figure 23.1.

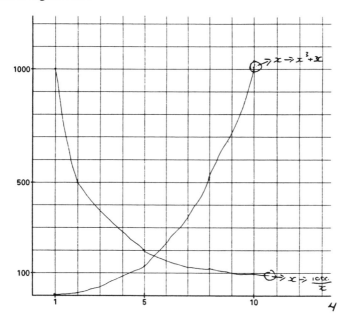

Fig. 23.1

Students gave a wide variety of answers to the last question:

- "Just solve the equation $x^3 + x = 1000/x$."
- "Take fractions and decimals instead of whole numbers."
- "Look at the graph and start substituting 5.6 and then 5.61, 5.62, 5.63, etc."
- "From the graph you know it's between 5 and 6. So start exploring this domain in more detail."

Readers may be surprised to learn that Dutch teachers were given almost no help in grading the examination, even on this last question. The grading instructions just say, "4 points for proper reasoning." The rest is left to the teacher, who is still regarded as being an expert in such matters.

Another more geometrical problem from the 1992 central examination was less succesful, although the teachers considered it fair and the level appropriate and students had encountered activities similar to this item in school.

Count Floris is on his way to conquer the city of Horn. He would like to know the plan of the city. For that reason he has sent out three spies. The spy who was looking at the city from the west made the...drawing [in fig. 23.2]. The spy who made a drawing from the northeast came up with [the drawing in fig. 23.3]. The third spy was looking at the city from the south, but the spy didn't return to the camp. "No problem," said count Floris, "we can make that drawing ourselves."

Horn vanuit het westen

Fig. 23.2

Horn vanuit het noordoosten

Fig. 23.3

On the worksheet [not shown here], you will see the circular wall of the city of Horn.

1. Show by making a drawing how you can properly place the three towers of Horn.
   - Put a *G* on the site of the castle on the map.
   - Put a *B* on the site of the dome on the map.
   - Put a *P* on the site of the tower on the map.

On the worksheet you see an incomplete drawing of the city from the south.

2. Draw the site of the three towers on the map.

Let us first look at a more or less successful answer (fig. 23.4).The complete worksheet is shown. The student has correctly drawn three parallel lines from the west (through points *B*, *P,* and *G*, respectively). Next he has drawn three lines from the northeast, using the sketch from that direction. The intersection of the lines at point *B* gives the position of *B* in the city, and in a similar way we can find the other towers.

Presumably the intention of the test designers was that students draw parallel lines from points *B*, *P,* and *G* on the map to the corresponding towers on the picture on the same worksheet of the view from south of the city. At one school we studied we noticed that none of the twenty students had done so. Actually, we are not certain that this action would have been correct because it is unclear if the drawings on the worksheet were related to each other in such a way that they could be used as one construction. We definitely need more experiences and careful analysis to find out what went wrong, to what extent, and why. Most students showed results that hinted at a better understanding than was visible in their work.

The teacher remains a key factor in the reform of testing. He or she has to accept wholeheartedly the changing emphasis on more open-ended, complex problems. Teaching will become more difficult and complex as well. The teacher may lose some authority when students present clever solutions. He or she will do less telling and will interact more with the students in the discussion

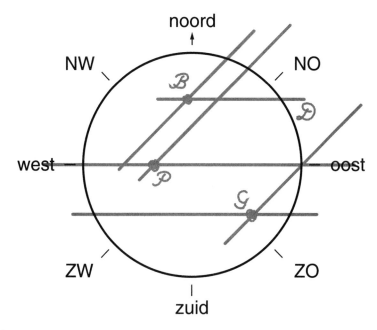

Vrang 17

Horn vanuit het zuiden

Fig. 23.4

of solutions. So even if the test producers succeed in making better achievement tests (in this instance, at the national level), the teacher remains the most critical factor. He or she deserves a lot of attention and help in designing tests. Fewer restrictions give the teacher a wide variety of possibilities in test design and administration. Thus test problems can be more exciting and rewarding but more difficult for the teacher to invent and grade.

SCHOOL TESTS

For testing achievement at the lowest level, the TRWT remains a very good tool. Even under time restrictions, students can perform well at higher levels, given the proper questions. The next example is taken from a fifty-minute school test that consisted of three problems. The sixteen-year-old students were taking the A curriculum.

Here you see a crossroads in Geldrop, the Netherlands, near the Great Church [fig. 23.5].

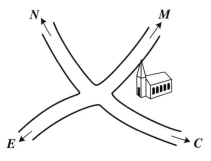

Fig. 23.5

To let the traffic flow as smoothly as possible, the traffic lights have been regulated so as to avoid rush-hour traffic jams. A count showed that the following number of vehicles in matrix A passed the crossroads each hour during rush hour:

$$
A = \text{From} \quad
\begin{array}{c}
\\ M \\ N \\ E \\ C
\end{array}
\begin{array}{c}
\overset{\displaystyle \text{To}}{\begin{array}{cccc} M & N & E & C \end{array}} \\
\left[\begin{array}{cccc}
0 & 40 & 200 & 30 \\
30 & 0 & 80 & 50 \\
210 & 60 & 0 & 60 \\
30 & 40 & 80 & 0
\end{array}\right]
\end{array}
$$

Matrices G_1, G_2, G_3, and G_4 [fig 23.6] show which directions have a green light and for how long. The entry "2/3" means that traffic can go through a green light for two-thirds minute.

- How many cars come from the direction of Eindhoven (E) during that one hour? How many travel toward the center of the city?

- How much time is needed for all the lights to turn green exactly once?

- Determine $G = G_1 + G_2 + G_3 + G_4$ and thereafter $T = 30G$. What do the elements of T signify?

- Ten cars per minute can pass through the green light. Show in a matrix the maximum number of cars that can pass in each direction in one hour.

- Compare your matrix to matrix A. Are the traffic lights regulated accurately? If not, can you make another matrix G that allows traffic to pass more smoothly?

$$
G_1 = \begin{array}{c} \\ M \\ N \\ E \\ C \end{array}
\begin{array}{cccc} M & N & E & C \\
\left[\begin{array}{cccc}
0 & \frac{2}{3} & \frac{2}{3} & 0 \\
0 & 0 & 0 & 0 \\
\frac{2}{3} & 0 & 0 & \frac{2}{3} \\
0 & 0 & 0 & 0
\end{array}\right] \end{array}
\qquad
G_2 = \begin{array}{c} \\ M \\ N \\ E \\ C \end{array}
\begin{array}{cccc} M & N & E & C \\
\left[\begin{array}{cccc}
0 & 0 & 0 & \frac{1}{3} \\
0 & 0 & 0 & 0 \\
0 & \frac{1}{3} & 0 & 0 \\
0 & 0 & 0 & 0
\end{array}\right] \end{array}
$$

$$
G_3 = \begin{array}{c} \\ M \\ N \\ E \\ C \end{array}
\begin{array}{cccc} M & N & E & C \\
\left[\begin{array}{cccc}
0 & 0 & 0 & 0 \\
0 & 0 & \frac{1}{2} & \frac{1}{2} \\
0 & 0 & 0 & 0 \\
\frac{1}{2} & \frac{1}{2} & 0 & 0
\end{array}\right] \end{array}
\qquad
G_4 = \begin{array}{c} \\ M \\ N \\ E \\ C \end{array}
\begin{array}{cccc} M & N & E & C \\
\left[\begin{array}{cccc}
0 & 0 & 0 & 0 \\
\frac{1}{2} & 0 & 0 & 0 \\
0 & 0 & 0 & 0 \\
0 & 0 & \frac{1}{2} & 0
\end{array}\right] \end{array}
$$

Fig. 23.6

Of course this excercise bears all the marks of a TRWT. It is relatively closed and has a series of short questions to guide the students. It would be interesting to find out what would have happened if we had posed only the last question.

School tests offer other exciting possibilities. Some of them were explored in the Netherlands and the United States in recent experiments.

Two-Stage Tasks

The first stage is carried out like a traditional timed written test. The students are expected to answer as many questions as possible within a fixed time limit. After having being graded by the teacher, the tests are handed back to the students with their scores. The second stage follows. With the benefit of information about their mistakes, students repeat their work at home without any restrictions, thus affording them the opportunity to reflect on their activities in the first stage of the task.

The Essay

In the American and Dutch experiments, a certain kind of essay question was very successful. The students were given an article from a magazine with a lot of information in tables and in numbers. The students were asked to rewrite the article by making use of graphical representations. The article discussed the problem of overpopulation in the Republic of Indonesia. One of the graphs designed by students shows clearly the large differences in the spread of the population over the islands (fig. 23.7). The left bar on each island shows the area of the island as a percent of the total area. The right bar shows the

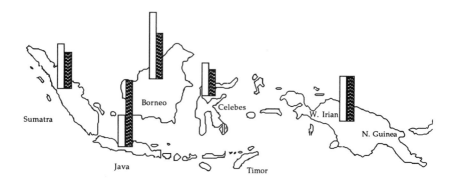

Fig. 23.7

population of the island as a percent of the total population. The graph clearly shows that Java is overpopulated and that more space is available on some of the other islands.

The "Test" Test

One of the more promising new ideas gives students the task of creating a test. They are given the following directions:

> This task is very simple. By now you have worked through the first two chapters of your book and taken a relatively ordinary test. This test is different. Design a test for your fellow students that covers the whole book. You can start your preparations now. Look in magazines, papers, books, and so on, for data, charts, and graphs that you want to use. Keep the following in mind:
>
> - Students should be able to complete your test in one hour.
> - You should know all the answers.

The results of such a test are surprising. Students are forced to reflect on their own learning process, and the teacher obtains an enormous amount of feedback on his or her teaching activities. It is too early to draw conclusions about this kind of testing, but later this year we will publish results from experiments at an American school. A ninth-grade girl studying data visualization produced the following test question. Developing a critical attitude was one of the general goals. (Refer to the graph in fig. 23.8.)

- What have the makers of this graph done to save space?
- What impression does this give the reader?
- What could be done to make the graph less deceiving?

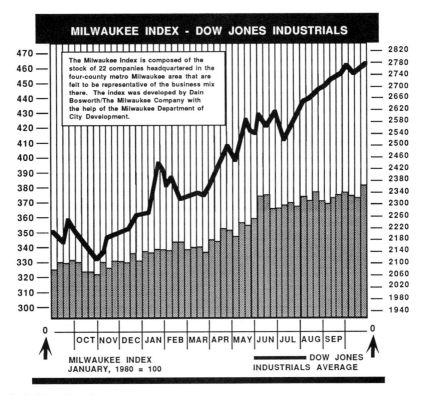

Fig. 23.8

Initial results of this form of testing indicate that the students are hampered by their traditional experiences of teaching and learning. They develop tests of lower skills that often conform to chapter divisions and do not integrate the topics studied. On the positive side, we noted that almost all exercises were different in context and subject and displayed common elements in noncontextual mathematics (see Lange et al. [1992]).

CONCLUSION

The evidence of these interesting developments is convincing. Open-ended questions requiring higher-level thinking promote conceptual mathematization. The barriers, however, are massive. We must abolish multiple choice once and for all. We must design clever open-question tests (TRWT) and come up with alternative ideas in testing that adhere to the mentioned principles. We must investigate the actual effects on learning and teaching of using appropriate tests and evaluate their practicality. We must encourage test developers to develop the proper tests and design an innovative strategy to convince teachers and (especially in the United States) parents and politicians of the merits of such testing.

REFERENCES

Cockroft, W. H. *Mathematics Counts: Report of the Commission of Inquiry into the Teaching of Mathematics in Schools.* London: Her Majesty's Stationery Office, 1982.

Freudenthal, Hans. *Didactical Phenomenology of Mathematical Structures.* Dordrecht, Netherlands: Reidel, 1983.

_____, *Revisiting Mathematics Education.* Dordrecht, Netherlands: Kluwer Academic Publishers, 1991.

Galbraith, P. L. "Paradigms, Problems and Assessment: Some Ideological Implications." In *Investigations into Assessment in Mathematics Education: An ICMI Study*, edited by Mogens Niss. Dordrecht, Netherlands: Kluwer Academic Publishers, 1992.

Gravemeijer, Koeno, Marja van den Heuvel, and Leen Streefland. *Contexts, Free Productions, Tests and Geometry in Realistic Mathematics Education.* Utrecht, Netherlands: Onderzoek Wiskundeonderwys an Onderwijs Computer Centrum (OW&OC), 1990.

Lange, Jan de. *Mathematics, Insight and Meaning.* Utrecht, Netherlands: OW&OC, 1987.

Lange, Jan de, M. van Reeuwijk, Gail Burrill, and Thomas Romberg. *Learning and Testing Mathematics in Context—the Case: Data Visualization.* Scotts Valley, Calif.: Wings for Learning, 1992.

Streefland, Leen. *Fractions in Realistic Mathematics Education.* Dordrecht, Netherlands: Kluwer Academic Publishers, 1991.

Treffers, Adri. *Three Dimensions: A Model of Goal and Theory Description in Mathematics Education.* Dordrecht, Netherlands: Reidel, 1987.

24

Assessment in the Interactive Mathematics Project

Lynne Alper
Dan Fendel
Sherry Fraser
Diane Resek

THE Interactive Mathematics Project (IMP) has created a problem-centered secondary school curriculum to fulfill the vision of NCTM's *Curriculum and Evaluation Standards* (1989). The IMP curriculum is designed to help change ideas about mathematics education in several key ways, including who should study mathematics, what mathematics ought to be taught, how mathematics is best learned, and how student learning can be assessed.

Opportunities for assessment of learning go beyond the traditional paper-and-pencil test to include self-assessment, student portfolios, oral presentations, written explanations, teacher observations, and group work.

The varied tools used for IMP assessments yield evaluation on several levels:

- Students learning about their own progress
- Teachers learning about the progress of individual students
- Teachers learning about their teaching and their curriculum

We concentrate primarily on the first two types of learning by looking at actual student work and how it can inform the student and the teacher. All the student work comes from the IMP unit "The Pit and the Pendulum," which we describe.

First, however, we should put this discussion in the context of some of the other basic assumptions and features of the project:

- IMP represents a shift from a skill-centered to a problem-centered curriculum.
- IMP represents a broadening of the scope of the secondary school

The authors thank the following students in the program whose work we have used under pseudonyms: Jacinda Abcarian, Deji Abina, Carrie Hunt, Nhut Nguyen, Micah Porter, Lisa Von Blankensee, and Zack Walter. We have transcribed their words verbatim, including punctuation and spelling errors, and copied graphics directly from their papers.

curriculum to include such areas as statistics, probability, and discrete mathematics.

- IMP represents changes in pedagogical strategies, including emphasis on communication and writing skills.

- IMP represents expansion of the pool of students who receive a "core" mathematics education.

"THE PIT AND THE PENDULUM"

The IMP curriculum is composed of four- to six-week units, with each unit focusing on a central problem. "The Pit and the Pendulum" comes in the second semester of the high school mathematics program. The unit opens with an excerpt from Edgar Allan Poe's story, in which a prisoner is tied down while a thirty-foot pendulum with a sharp blade slowly swings over him. If the prisoner does not act, he will be killed by the pendulum. When the pendulum has about twelve swings left, the prisoner creates a plan for escape and executes it. Students are presented with the problem of whether the prisoner would have time to conceive of and execute a plan of escape. Students see that the key question is, How long would twelve swings of a thirty-foot pendulum take? To resolve this question, students construct smaller pendulums and conduct experiments to find out what variables determine the length of the period of a pendulum and what the relationship is between the period and these variables.

In the course of the unit, students make and refine conjectures, analyze data collected from experiments, learn about and apply ideas on standard deviation to tell whether a change in one variable affects another, study quadratic equations from the Green Globs and Graphing Equations (Dugdale and Kibby 1986) software program, and use graphing calculators to explore curve fitting. Finally, after deriving a theoretical answer based on their data, students build a thirty-foot pendulum to test their theory.

STUDENTS LEARNING ABOUT THEMSELVES

When students feel that what they are studying is important and that they are really learning, they are much more likely to put forth effort in school. Thus, students must be given time to reflect on their progress. The reflection time is also helpful to them in learning which of their behaviors tend to facilitate learning and which inhibit their learning. Through reflection, students become aware of which skills and understandings they need to improve and acquire.

The IMP curriculum provides this reflection time in several ways, of which we present two contexts in this article:

- Students writing the cover letter for their unit portfolios
- Students describing their process in solving the Problem of the Week

Portfolios

At the end of each unit, students compile a summary portfolio by choosing quality samples of their work that reflect the mathematics in the unit. These samples include homework papers, reports on classroom group investigations, and write-ups on Problems of the Week, all of which have been previously assessed in some manner and, in many instances, graded.

After reviewing their work and selecting materials for their portfolios, students write a cover letter, as described in figure 24.1. Students know that at any time during the year, they, their parents, or educators can read through their portfolio and assess progress in communication skills and mathematics concepts.

Cover Letter

Write a cover letter to go with your portfolio. Include in your cover letter—

1. a brief description of what the unit was all about;
2. what mathematics topics you studied;
3. what you learned;
4. why you chose each of the items for your portfolio;
5. how you think that you progressed in these areas:

 • Working together with others
 • Presenting to the class
 • Writing about and describing your thought processes

6. which areas you need to work on:

 • Mathematics topics
 • Other classroom processes

7. any other thoughts you might like to share with a reader of your portfolio.

Fig. 24.1. Instructions for the cover letter for students' portfolios

Carla's letter begins with a summary of the unit's content and then comments on the use of the graphing calculator. She writes,

> The lessons on calculators, and how to use them, was somewhat confusing, since there were so many buttons and things to remember. The slightest mistake, like pressing one wrong button, could totally mess up one's whole program.

Carla will probably be very careful in the future when pushing calculator buttons. Her complaint also suggests that the teacher should stress the overall purpose of the calculator activity, lest students get bogged down in details.

Carla then comments,

> The groupwork made all our assignments get done easier and quicker. I feel I need
> to seriously work on getting my assignments done ON TIME, because if you leave
> one assignment out, then another, it really builds up, and isn't easy to do all at one
> time. That is what I regret this semester (not doing assignments on time).

Carla has learned valuable lessons about the sharing of tasks as well as the
importance of continuity and keeping up with the learning in the unit.

Problems of the Week (POWs)

To give students experience in working independently on substantive
problems, IMP has incorporated Problems of the Week (POWs) into each unit.
Some POWs are directly related to the specific mathematics of the unit under
study; others are classic mathematics problems, independent of the unit, that
require students to use logical reasoning in a variety of problem-solving
contexts.

A POW from "The Pit and the Pendulum" (fig. 24.2) is of the latter type. The
reader should take some time to work on "Eight Bags of Gold" so as to
understand better the student work that follows.

When writing a description of the process they use to reach a solution,
students become aware of what worked for them and what didn't in this
problem situation. Perhaps what worked last week will not work this week. This
time for reflection helps students choose promising strategies for new problems.
Here is part of Jorge's write-up:

> *II. Process:* I could not get started on this problem because I was not sure what the
> "fewest number of weighings" ment. Did it mean the least possible # of weighings
> or did it mean least # of weighings provided he did not want to take a chance and
> luckly put one bag on each side of the scale and find it on his first try. So I had to
> assume the king wanted to not take any chances on luckly picking the right one.
> This ment I had to find a system to way the bags were he would not make any
> lucky guesses. I got started by drawing different ways the king could have done
> the weighings.

Jorge seems to have learned that clarifying the problem is a good first step.

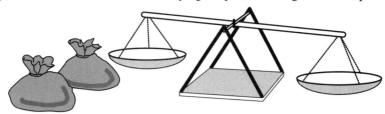

Cornell does a good job of tracing his progress toward a solution through
several approaches that didn't work.

POW #12—Eight Bags of Gold

Once upon a time there was a very economical king who gathered up all the gold in his land and put it in 8 bags. He made sure each bag weighed exactly the same amount. He chose the 8 people in his country that he trusted the most, and gave a bag of gold to each of them to keep safe for him. On special occasions he asked them to bring the bags so he could look at them. (He liked looking at his gold even if he didn't like spending it.)

One day the king heard from a foreign trader that someone from the king's country had given the trader some gold in exchange for some merchandise. The trader couldn't describe the person who had given her the gold, but she could tell it was someone from the king's country. Since the king owned all the gold in his country, he knew that one of the 8 people he trusted was cheating him.

He decided to ask the 8 trusted people to bring their gold to him. The only scale in the country was a pan balance. This wouldn't tell how much something weighed, but it could compare two things, and tell which was heavier and which was lighter. The person whose bag was *the lightest* would clearly be the cheat.

Being very economical, the king wanted to use the pan balance as few times as possible. He thought he would have to use it three times in order to be sure which bag was lightest. His wizard thought that it could be done in fewer weighings. What do you think?

- What's the fewest number of weighings he would need?
- How can you be sure?

Note: If you make a comparison, and then take one or more bags off each side, it counts as a new weighing.

Write-up

1. *Problem statement:* In your own words, state the problem clearly enough so that someone unfamiliar with the problem who picked up your paper could understand what it is you are asked to do.

2. *Process:* Based on your notes, describe what you did in attempting to solve this problem:
   - How did you get started?
   - What approaches did you try?
   - Where did you get stuck?
   - What drawings did you use?

   Include things that didn't work out or that seemed like a waste of time. Do this part of the write-up even if you didn't solve the problem.

3. *Solution:*
   - State your solution(s) as clearly as you can. (If you only obtained a

partial solution, give that. If you were able to generalize the problem, include your more general results.)

- Explain how you know that your solution is the best one possible. Your explanation should be written in a way that will be convincing to someone else—even someone who initially disagrees with your "answer." Remember that merely stating the answer will count for nothing!

4. *Extensions:* Invent some extensions to this problem. That is, write down some questions that might be related to this problem. They can be easier, harder, or about the same level of difficulty as the original problem. (You are not expected to solve or answer these additional problems or questions.)

5. *Evaluation:* Was this problem too easy, too hard, or about right? Explain why.

Fig. 24.2. POW #12—"Eight Bags of Gold"

2) Process: This problem at first seems easy then it gets hard, then easy. First off I thought that 8 bags of gold I should draw a table like this:

	1	2	3	4	5	6	7	*light one* 8	
1	🛡	𝓔	𝓔	𝓔	𝓔	𝓔	𝓔	𝓤	
2	𝓔	🛡	𝓔	𝓔	𝓔	𝓔	𝓔	𝓤	𝓔 = *Weigh the same*
3	𝓔	𝓔	🛡	𝓔	𝓔	𝓔	𝓔	𝓤	
4	𝓔	𝓔	𝓔	🛡	𝓔	𝓔	𝓔	𝓤	𝓤 = *Uneven (one weighs less)*
5	𝓔	𝓔	𝓔	𝓔	🛡	𝓔	𝓔	𝓤	
6	𝓔	𝓔	𝓔	𝓔	𝓔	🛡	𝓔	𝓤	
7	𝓔	𝓔	𝓔	𝓔	𝓔	𝓔	🛡	𝓤	
light one 8	𝓤	𝓤	𝓤	𝓤	𝓤	𝓤	𝓤	🛡	

This didn't seem to get me anywhere so I thought of weighing a pair at a time and that took 4 weighings. So I threw that away. And began to think about it, I then tried weighing 2 at a time on each side but that came up with three tries. Then I finally realized weighing three on each side, then weighing the other two. I knew I had to be Right.

Cornell seems to be learning an important lesson about "doing" mathematics: one often finds an approach that works only after trying many that don't work.

Take-Home Assessment

At the end of each unit, students have a two-stage assessment. The in-class portion is designed so that all students should be able to finish it during the allotted class period. Students who finish early may begin the take-home portion of the unit assessment.

No time limit is placed on the take-home part of the assessment. Students are allowed to get outside help as long as they acknowledge it in their write-up. Few actually seek such assistance. They are free to approach the problem in many ways. The way in which they incorporate the concepts of the unit into their responses helps the teacher assess what mathematics they were able to transfer to a new situation.

During "The Pit and the Pendulum," students had experienced conducting experiments and determining which variables affected the results. In the take-home assessment shown in figure 24.3, they are asked to *devise* an experiment. We do not focus on their use of arithmetic or on their numerical answer. We want to assess whether students can put the question asked in the assessment into the context of the mathematics studied in the unit.

Take-home assessment for "The Pit and the Pendulum"

A circus performer wants to ride a bicycle right up to a brick wall and stop very close to it without crashing. She wants to know when to put on the brakes. But she doesn't want to try it out, since she is afraid of crashing. She just wants to predict at what point she should apply her brakes.

Devise a plan to collect and analyze data that will allow her to make this prediction. Talk about the variables to consider, the problems she will encounter, normal distribution and standard deviation.

(You may use graph paper, notes from previous work, etc., in working on this assignment. While still in class, you may also use graphing calculators.)

Fig. 24.3

Here is part of Nevi's work:

MY PLAN:

Have her draw a line on the ground. This will be her brick wall. Have her do it on the same ground she will be using in the performance. Also have her use the same bike that she will use later. Then put 6 or 7 pilons in a line next to her bike-way. See ILL. Place them about a 1/2 foot apart going all the way to the wall. Start testing to see which distance (marked

by the cones) is best. Ride at an *even* steady pace, testing the closest one to the "wall" first & getting more distance if you need it. Allways apply brakes as hard as they will go. Once you get your best distance, double check it to make sure it's best.

Nevi shows that she understands the concept of a controlled experiment, but she does not seem to have understood that the measurements of the same data are not identical and are normally distributed.

Cori begins her write-up by creating some fictitious experimental data showing stopping distances of 16, 14, 15, 11, 16, 14, 19, 15, 13, and 17 feet and finding that the mean is 15 feet. She then continues as follows:

That means that she'd have to theoretically start to put on the brakes 15 ft. before the wall. In an ideal situation, the data you got would form a bell shaped curve:

But, because you have an equal amount of #'s on each side, some too big and some too small, half the time you would hit the wall, and half the time you would stop too far back. That would mean that the wall was right in the middle.

Well, that's just not good enough. You need to move the wall over to the end. Now you need to figure the standard deviation.

Cori proceeds to do so, finding that "$1\sigma = 2.1$, $2\sigma = 4.2$," and then explains how to use these numbers:

We know that if the wall is 2σ over from the mean, 95% will be before the wall. That's not quite true.

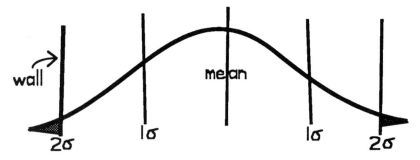

The 2 shaded areas on the ends are the parts that are outside of + or – 2σ's. Those areas equal 5%. But we only moved the wall one way; we moved it 2σ's to the left, so, the shaded part on the right is included in the room the performer has to ride without hitting the wall. That total space is equal to 97.5% (95 + 2.5), not 95%....

So, if the performer always travels the same speed, she can hit the brakes ... and still only have a 2.5% chance of hitting the wall.

Now all the circus performer has to do is to do her own tests, and do to them what I did to my "results". I hope you understood all of this.

Cori's work shows her thorough understanding of both the mechanics and the ideas involved in standard deviations.

In-Class Assessment

Certainly, teachers learn about the effect of their teaching and their curriculum by looking at individual students' work. But by looking at the work of a number of individuals, the teacher can assess her or his own teaching and curriculum. By seeing the strengths and weaknesses of the members of the class, the teacher can plan how to continue working with the present class and how to modify the curriculum and teaching strategies the next time she or he teaches the course.

The in-class assessment for "The Pit and the Pendulum" is shown in figure 24.4, followed by some student write-ups. This problem enables the teacher to observe and assess students' understandings and misconceptions about the concepts studied in the unit. The reader is advised to work on the problem before looking at the student work.

In-class assessment for "The Pit and the Pendulum"

Al works in a hobby shop. He  noticed the relation between the length of certain models and the amount of paint needed for them. Here are some figures he came up with:

Length of Model (in inches)	Amount of Paint (in milliliters)
1	4
2	12
3	28
5	80

Assuming that this pattern continues, predict how much paint would be needed for a model that was ten inches long. Explain your reasoning.

(You may use graphing calculators, graph paper, notes from previous work, etc., in working on this assignment.)

Fig. 24.4

Yas's work on the in-class assessment is interesting. As can be seen in his graph (fig. 24.5), he plots the given points on graph paper but accidentally plots the fourth as (5, 30) instead of (5, 80). He then makes his prediction fit the faulty data! Interestingly, his choice of scale makes it impossible for the fourth point and the one in question to fit on the paper. He needs to realize that choosing a scale is an important step that should come before he begins to plot his points.

> This assessment is almost like the one we had to find for the time of the thirty foot pendulum. What I did was first, I graph the numbers they gave me on the sheet of paper. Then I connected all the dots. When I was done during that, I predict that the line will make a right turn and keep increasing. I drew a line from the 10 inches up. From the line I predicted to the line I just drew, when the lines connected that will be the answer.

THIS IS WRONG

I shouldn't make a right turn

But notice that Yas realizes that he is in error. He recognizes that this graph doesn't look like those he has encountered before. Unfortunately, he is working only with the graph rather than also looking for a relationship between the numbers, that is, between the length of the model and the amount of paint needed.

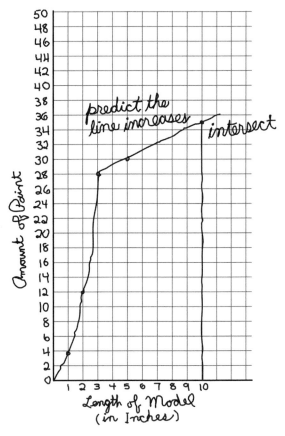

Fig. 24.5. Yas's graph

In our final example, Howard seems to have the key ideas of curve fitting well in hand. He uses the graphing calculator in several ways. He solves the problem by first plotting the given points and recognizing from the general shape that the function is quadratic. He then does some curve fitting until he is satisfied. Given a formula, he can then predict the result for one foot. (He was responding to a slightly different version of the problem.)

I made a graph and drew the curve of the points I had plotted. Then I plotted the points on the [graphing calculator]. I new from previous work that the equations for the graph would be something like $Y = X^2$ so I did this. The line was further away from the Y axis than the plotted points, so I decided to put $Y = X^2 \div 2$, this just made it go even further so I put $Y = X^2 \cdot 2$ this line was closer than any of the others but still not there so I kept adding a little bit like $Y = X^2 \cdot 2.2$, 2.4 ... ect. until I got to put $Y = X^2 \cdot 3$ this hid all of the points I figured it was the equation. Finally I plugged 12 (1 foot) in and solved the equation put $Y = 12^2 \cdot 3(Y = 432)$

If Howard's work is typical of the class, the teacher can comfortably move on, perhaps giving Yas some extra attention. But if Yas's paper is more representative, the teacher could conclude that the class needs more work on associating the algebraic relationship between the x- and y-coordinates with the graph, since this idea is deep and important.

REFERENCES

National Council of Teachers of Mathematics. *Curriculum and Evaluation Standards for School Mathematics.* Reston, Va.: The Council, 1989.

Dugdale, Sharon, and David Kibbey. Green Globs and Graphing Equations. Scotts Valley, Calif.: Wings for Learning/Sunburst, 1986.

25

The Classroom Assessment in Mathematics Network

Maria Santos
Mark Driscoll
Diane Briars

THERE is ferment and activity in the reform of assessment practices in mathematics across the nation. Many assessment projects are trying to capture what students know and can do in mathematics. One such project is the Classroom Assessment in Mathematics (CAM) Network, which strives to create a knowledge base on assessment within the context of urban schools.

The CAM project set itself the goal of structuring classroom experimentation around a series of questions. The questions were seemingly simple ones: What do students know? How can we best assess what they know? How can assessment improve instruction? The responses we are receiving, however, are far from simple, far more than expected, and have far-reaching implications for the future of mathematics education in urban centers.

The Classroom Assessment in Mathematics Network project brings together middle-grade teachers and mathematics supervisors from the Urban Mathematics Collaborative as well as staff from the Education Development Center in Newton, Massachusetts. The CAM project supports urban middle-grade mathematics teachers and supervisors in building a knowledge base in classroom assessment by working with them to review the theory and practice of classroom assessment and to support classroom experimentation. The Education Development Center facilitates the work of the participants by brokering resources and sustaining a critical dialogue on the CAM electronic network. The CAM electronic network enables teachers to share a task with their colleagues around the country, receive critical comments, and then revise the task.

A group of four middle grade teachers at each of seven urban centers are currently engaged in structured experiments in classroom assessment. The experimentation is structured to provide teachers with ample time for the design of tasks, the review of theory, the discussion of results, and reflection on classroom implications. Pairs of teachers at targeted schools have a common free period for daily meetings. They also meet twice a month with the local team, which is composed of four teachers and the mathematics supervisor. Open-ended questions, performance tasks, portfolios, investigations, and projects are being used in the experimentation. We are examining and refining these types of assessment tasks, asking ourselves, "What next?" We hope to expose in this article some of the early results from a collection of structured experiments in classroom assessment. The context of the work will be described, and some of the emerging issues will be explained.

EARLY RESULTS

Teachers in Cleveland, Dayton, Memphis, Milwaukee, Pittsburgh, San Diego, and San Francisco are hammering away at the complex issues that affect assessment in urban settings. As teachers have crafted ways in which to gain a fuller understanding of students' thinking and knowledge in mathematics, three typically urban issues have surfaced: diverse background knowledge, different values and belief systems, and communication. Teachers are finding that these three major urban issues need to be addressed if we are to gain full insight into students' understandings in mathematics.

Diverse Background Knowledge

The complexity of changing assessment practices in urban centers is magnified by the rich diversity of the urban classroom. Students from all socioeconomic backgrounds sit in the same classroom along with large numbers of students who are still developing proficiency in the English language. Newcomer multilingual and multicultural students share the classroom with third- and fourth-generation members of various ethnic groups. Transient student populations of migrant workers, gypsies, and the homeless enter, leave, and return to the system regularly. The rich tapestry of the urban classroom provides both challenges and opportunities to generate multiple points of entry for all students.

Teachers in urban classrooms using context-enriched assessment items have found that the assessment tasks fail to produce sufficient knowledge about their students' thinking and understanding because culture-rich context limits entry for many students. A culture-rich context is one that embeds experiences and values that are native to particular groups of students. Students from diverse backgrounds do not share the same type of experiences and opportunities. Their home life is very different from that of mainstream America. Items, tools, expressions, issues, and experiences common to others may not exist in their world. When faced with a context-enriched assessment task, these students must first understand the context. When using context-enriched assessment, teachers have first had to prepare the students by exposing them to the context through preassessment activities.

At times teachers present tasks to students that seem appropriate and even relevant to their lives, only to find out quickly that there is little or no point of reference for understanding the context of the task. The task then becomes inaccessible. Sandy from San Francisco posed the problem depicted in figure 25.1.

Sandy quickly found that many of his students didn't have any idea of what a matte was, but many were intrigued by its use. Sandy brought several mounted pictures to class to heighten his students' understanding of the context of the problem. If he uses the problem in the future, he will provide preactivities on the matte, pictures, and framing. Teachers in other sites have had to address issues involving "time" and "candles" with students having diverse backgrounds.

Value and Belief Systems

Structured experimentation has increased teachers' awareness of differences between what they value and what their students seem to value. When opening up assessment tasks, the teachers have enhanced or modified tasks to create an interesting situation, multiple entry points, multiple solutions, and an audience. The new tasks require students to explain their thinking, provide the most appropriate representation, and justify their responses. This calls for students to do more than explain a sequence of steps or procedures in looking for a resolution. This type of assessment asks students not only to apply mathematics but also to reflect on the application of mathematics and the mathematics itself. Through the experimentation, teachers have become aware of the difficulties this creates for many urban students. Many students have been in classroom settings and in homes that do not ask them to think, analyze, argue, plan, and revise when trying to solve a problem.

The tension is magnified as teachers find themselves questioning their own belief systems both in the teaching of mathematics and in their expectations of students. The transition from a belief system that has at its center the goal that students remember what is told or shown toward one where what is valued is critical thinking and the construction of new knowledge by students is daunting. The struggle between the belief systems is exemplified in the tasks asked of students. At times a traditional question is opened up, but the new task reflects the traditional values of remember and show.

One teacher developed this task: "A carpenter needs to cut two pieces of lumber, 15 1/4 in. long and 15 5/8 in. long, out of a piece of wood that is 30 3/4 in. long. Is the larger piece of wood long enough? Explain your answer in words and diagrams." Students from four different teachers responded to the task, and a discussion of the task was held by the group of teachers and the supervisor. Some students were successful in drawing and explaining in words how they

FRAMING A PHOTOGRAPH

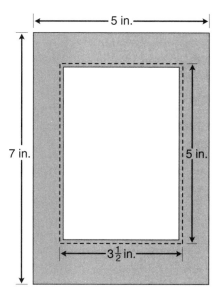

You want to frame a picture of yourself to give to your best friend. The photograph is 3 1/2 in. by 5 in. and is indicated by the dashed lines in the drawing above. Your frame is 5 in. by 7 in. The person who is cutting the matte (border) for your picture is just learning how to do it, so you have to give him or her very explicit instructions.

Give complete and exact instructions about how to measure and cut the rectangular opening in the center of the matte. The photo must be overlapped by the matte by 1/8 in. to 1/4 in. all the way around. Assume the framer has a ruler that measures to 1/16 in. accuracy. The opening for the photo must be perfectly centered in the frame. Give all necessary dimensions, and suggest a technique to be sure that all cuts are parallel to the sides of the frame. Use inches and fractions of inches in your instructions.

Make a true scale drawing of how to cut the matte board with all dimensions as well as written instructions. Your drawing should look something like the drawing above. However, the drawing above is not exactly to scale. Give your solution on a separate piece of paper. Use a ruler and be neat. Be sure to explain how you came up with the numbers you used for all dimensions. If you say, "I guessed this number, and it worked," that will not be acceptable.

Fig. 25.1. An example of a culture-rich context for a task

solved the problem, some students just used algorithms, and a couple of students questioned the task. The students who questioned the task felt that it was redundant both to explain in words and to use a diagram. In the discussion among the teachers, the question was, "What is the objective of the task? What are we testing for?" The teachers agreed that what they wanted to ascertain was students' sense of number. If we truly wanted to understand students' number sense, then the question should have been, "Explain to the carpenter the most efficient way of doing the task and justify your solution." The first rendition of the carpenter task placed the students somewhere in between "remembering and showing" and "critical thinking and construction of knowledge." This may be a developmental stage for teachers as they change their belief systems.

How do we change teachers' beliefs and values? In San Francisco, case studies developed by participating teachers have been used in the structured experimentation to examine and gain an understanding of the results. A case study is the description of a classroom scenario that depicts a delemma or critical situation based on a teacher's recollection. The case study shown in figure 25.2 exemplifies the difficulty of negotiating belief/value systems.

After the teacher presented her case study, she had three questions for the group:

- What can I do to aid the transition from manipulatives to pencil and paper?

- What can I do to help my students visualize fractions other than use the basic pattern blocks of 1/2, 1/3, 1/6?

- Is it a major problem that some children are not making the transition from pattern blocks to paper and pencil?

The group used the teacher's questions as a springboard for discussion. The discussion quickly centered on the child's misconceptions in division. The way in which this child learned division did not help distinguish that the remainder was not a whole number but a fraction or ratio. This misconception was interfering with the child's understanding of simplification. The discussion continued to suggest that elementary school teachers should not use the word *remainder* and that division should be taught in context and not as an isolated algorithm.

Communication

These new assessment tasks are asking students to communicate in many ways. Students are required to explain their thought processes, to justify their answers, to revise their work, and to write to an audience. Getting students to communicate effectively is very difficult when for some students the mathematics class is a safe haven from the world of writing. For many students still developing the English language, mathematics class is a place to excel or where language is more neutral. Explaining their ideas in English and justifying

Setting: Sixth-grade heterogeneous class of 28 students.

In February, we began with pattern blocks to teach fraction concepts: the yellow hexagon = 1; therefore, the red trapezoid = 1/2, the blue trapezoid = 1/3, and the green trapezoid = 1/6. We used these to add, subtract, and change denominators (looking for like denominators). As a check for understanding, I used other forms for 1 (such as two yellow hexagons = 1; therefore one red trapezoid = 1/4; or a yellow hexagon + a red trapezoid = 1, and so one red trapezoid = 1/3, and so on). By the end of March, the children seemed competent in adding and subtracting fractions with like and unlike denominators using the pattern blocks and in working with simple fractions (1/2, 1/3, 1/4, 1/6) using only pencil-and-paper, with pattern blocks available for back-up. Classwork and homework during these months included pencil-and-paper manipulations and writing short explanations of what they were doing with the pattern blocks.

We then did not have math class for about three weeks because of spring break, CTBS testing, and camping.

Yesterday, I gave the following problem as a warm-up exercise:

$$
\begin{array}{r}
4 \quad 3/5 \\
3 \quad 2/5 \\
+\,7 \quad 2/5 \\
\hline
\end{array}
$$

A child volunteered to share her answers:

$$
\begin{array}{r}
4 \quad 3/5 \\
3 \quad 2/5 \\
+\,7 \quad 2/5 \\
\hline
14 \quad 7/5 = 16\ 1/5
\end{array}
$$

I questioned her on the simplification. She said, "Seven-fifths is an improper fraction, so I divided five into seven. I got a remainder of 2 so I carried it over to the whole numbers. The 1 I put in the numerator. That's how I got 16 1/5." This child was very confident she was correct and had followed all the correct procedures. Notice she used mathematical terms. I tried to demonstrate what she should have said but feel I didn't do a good job. Certainly, she looked unconvinced at the end of the lesson. Part of my problem is that we had never dealt with fifths with pattern blocks. I drew seven triangles on the board and then asked, "How many make one whole?" She answered "5" and I circled them and asked what remained. "2." (I sensed she was impatient because she'd already told me this.) I redrew the five triangles as a pentagon with two sides left over and stated, "The five equal one whole." She looked without comprehending. I went back to the pattern blocks and pulled out seven green triangles. I asked her how many triangles make a whole? (6) How many are left? (1) So seven triangles equals what? (1 1/6) She had no problem with that exercise, but I feel she remained unconvinced that her method on the review problem was incorrect.

Fig. 25.2. A case study

their answers is foreign to most students. Teachers have witnessed a significant amount of resistance from students. They have also had to learn from their language arts colleagues ways in which to engage students in writing and have begun to develop strategies specific to mathematics. Assessment is becoming a reflective tool for students.

Rosann from Milwaukee had finished a unit on data collection and data analysis. She wanted to find out if students who had not done a similar problem could pool their experiences and "solve" a bar graph. She posed the problem in figure 25.3 to her students.

When Rosann examined the results of the students' work, she found that—

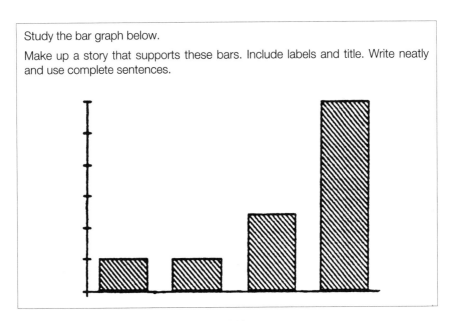

Study the bar graph below.
Make up a story that supports these bars. Include labels and title. Write neatly and use complete sentences.

Fig. 25.3

- 27 out of 97, or 27.8 percent, did not write a story, although they did complete the labels and title;

- 63 out of 97, or 64.9 percent, made a mistake with the labels or titles.

Some students' work exemplified the need for more writing, as illustrated in figures 25.4 and 25.5.

To improve the assessment, she is thinking of discussing a number of students' stories using an overhead to model expectations. She would also have students revise their work to ensure the use of proper labeling. Providing a menu of different types of graphs will help students improve their understanding of data collection and analysis Rosann states, "Students need to write more. It really was interesting to read what they wrote! No two papers were the same."

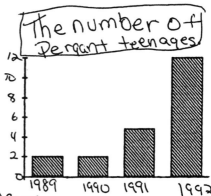

The first two years
it been the same average
But, 1991 went up 1992 thats the
nighes rate their is.

Fig. 25.4. One student's story about a bar graph

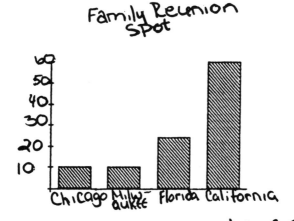

We are deceding were to have our
next family reunion. The family
is not too cooperative so we voted
one the main ones. The bar
graph shows our conclusion

Fig. 25.5. A second students's response about a bar graph

Mark from Pittsburgh has been working on getting his students to write. A major issue is communicating to students what a good response looks like. Mark has been using a strategy that seems to be working well—he asked students to work on a problem like this one:

> You and your friend are cutting Mr. S's grass. You get there early and cut the front yard alone. Both of you cut the back yard together. If the front yard and the back yard are the same size and Mr. S is paying you $10, how much should each of you get and why?

Then he made transparencies of a number of different solutions and asked students to comment on what they liked about each one (Mark used poor as well as good solutions as examples). After the students had seen a number of solutions, he asked them to describe the features or characteristics of a good response. They came up with the following criteria:

- It must show the math computation.
- It should have a diagram.
- It must describe how you got your answer (in words).
- It must have the right answer.

WHAT NEXT?

The work the CAM project is doing has just begun to bring to the surface some of the complex issues urban teachers are trying to address through instruction and assessment. These are issues that *can* be managed in the classroom and that *can* provide all urban students access to mathematics. They are issues like violence, indifference, humiliation, poor nutrition, and others that affect the daily lives of urban students, but they are issues on which educators in urban centers can focus attention so that more urban students enjoy and learn mathematics.

26

Student Self-Assessment in Mathematics

Patricia Ann Kenney
Edward A. Silver

I N *Thinking through Mathematics,* Silver, Kilpatrick, and Schlesinger (1990) assert that "we cannot retain much of the mathematics we have seen or heard if we have not appropriated it as our own" (p. 21). In order to appropriate mathematics as their own, students must assume an active role in their own learning by becoming aware of what they know about mathematics and by being able to evaluate their attainment of mathematical power. The process of actively monitoring one's own progress in learning and understanding and of examining one's own mathematical knowledge, processes, and attitudes is often called *self-assessment.* This article discusses student self-assessment from two different, yet related perspectives: (1) how students can use, and benefit from, self-assessment activities in mathematics and (2) how mathematics teachers can foster student self-assessment opportunities in the classroom and benefit from information obtained from such activities.

THE SELF-ASSESSMENT PROCESS

Before attempting to discuss the ways in which student self-assessment can benefit both students and teachers, it may be useful to examine certain aspects of the self-assessment process. Self-assessment is closely tied to metacognition, which refers to one's knowledge concerning one's own cognitive processes and products (Flavell 1976, 1979). Researchers in mathematics education (e.g., Garofalo and Lester 1985; Lester 1988; Schoenfeld 1987; Silver 1985; Silver, Branca, and Adams 1980) as well as in psychology (e.g., Brown 1978; Brown et al. 1983) have studied some important aspects of the metacognitive process. Of

The preparation of this article was supported, in part, by the Ford Foundation through Grant No. 890-0572. The opinions expressed are those of the authors and not necessarily those of the Ford Foundation.

these aspects, two are particularly relevant to this discussion of student self-assessment: being aware of one's own personal storehouse of content knowledge and cognitive processes (knowing about what you know) and engaging in self-evaluation (regulating and monitoring what you are doing while you are doing it).

Self-awareness, the first of these two components of self-assessment, often involves taking stock of one's own repertoire of mathematical knowledge, processes, strategies, and attitudes. According to Campione, Brown, and Connell (1988), a particularly powerful technique that can be used to operationalize self-awareness involves students actually describing or stating what they think they know about themselves as learners and about the resources they have at their disposal. Many opportunities exist in the mathematics classroom for students to engage in self-awareness. For example, while working on a solution to a problem, students might record whether they read the whole problem before beginning the solution process, whether they had seen a similar problem, or whether they used a particular solution strategy. During group work, students might document their role in the group (setting up equipment, recording observations) or their level of participation (listening to the others, serving as group leader).

Self-evaluation, the other important component of self-assessment, involves going beyond mere self-awareness and taking a critical look at one's own mathematical knowledge, processes, and disposition. Verbs such as *monitor, regulate, reflect,* and *oversee* are often associated with self-evaluation. During a problem-solving activity in the mathematics classroom, students who engage in self-evaluation step back from their work to think about how effective their strategies actually were or to examine the wisdom of continuing with their original plan for a solution. In their self-evaluation of their activities in cooperative groups, students think about their role in the group and reflect on how this role has contributed to (or detracted from) the group process and product.

In practice, the self-awareness and self-evaluation components of student self-assessment need to occur in tandem. Conscious knowledge about the resources available to them and the ability to engage in self-monitoring and self-regulation are important characteristics of self-assessment that successful learners use to promote ownership of learning and independence of thought.

STUDENTS AND STUDENT SELF-ASSESSMENT

Self-assessment is a common activity for many people. For example, amateur and professional athletes often evaluate their own performances through such activities as watching videotapes. By observing themselves in action and reflecting on their performance, athletes can focus on their strong points and work to maintain them; when their weaknesses become apparent, they can take steps to eliminate them. Similarly, by engaging in activities that involve self-awareness and self-evaluation of their mathematical performance, mathematics students can enhance their strengths and reduce their weaknesses.

The extent of students' awareness of their strengths and weaknesses is known to be associated with their success or lack of success in some areas of mathematical performance. For example, the literature on mathematical problem solving (e.g., Campione, Brown, and Connell 1988; Krutetskii 1976; Schoenfeld 1987) is replete with descriptions of successful problem solvers as students who have a collection of powerful strategies available to them and who can reflect on their problem-solving activities effectively and efficiently. In contrast, descriptions of unsuccessful problem solvers tend to portray them as students who have command of fewer strategies and who do not function in a self-reflective or self-evaluative manner.

Putting It in Writing

As discussed previously, students describing or stating what they think they know encourages self-awareness of the intellectual content of their activity and the information or techniques they have at their disposal. Examples of self-awareness through written activities can be found in a variety of sources (e.g., Campione, Brown, and Connell 1988; Resnick 1989; Stenmark 1991). Some common methods that encourage students to write about what they know about mathematics and about themselves as learners include filling in grids or charts, completing sentence starters, or keeping a mathematics journal. By encouraging students to "put it in writing," a teacher can remind students to take "time out" for self-assessment. Whatever recording method that students choose or that their mathematics teacher suggests, writing about their knowledge and strategies can motivate them to engage in self-assessment, as illustrated in the following scenario.

As part of a written homework assignment for her precalculus class, a student was asked to "describe a real-world example of a step function that is not in your textbook, try to explain in words why you think it is a step function, and sketch the graph." (This step-function example was adapted from a classroom activity used by Cathy Schloemer, a mathematics teacher at Indiana Senior High School in Indiana, Pa.) Because at first this student could not think of a good example other than the parking-lot example in her textbook, she decided to organize her thoughts, using a chart that her teacher had suggested. Figure 26.1 shows her completed chart for the step-function example. By cataloging the knowledge she already had (a correct graph of a step function), analyzing the goal (find a real-world example), and writing down some ideas (price relative to weight, postage), this student used self-assessment strategies to assist her in thinking through the assignment.

What I know	What I need to know	Think space
What a step function looks like $f(x) = [x]$, *greatest integer function*	*A real world example (other than parking lot:* 0-1 hrs $2.00 1-2 hrs $3.00 2-3 hrs $4.00 etc.)	*things that behave this way:* *- level,* *- then jump,* *- stay level* *- then jump again* *Price relative to weight?* *Postage?* *29¢ for so many oz, then 23¢ more. Call Post Office! This will work.*

Fig. 26.1. Step-function chart (adapted from Campione, Brown, and Connell [1988] and Resnick [1989])

Asking Self-monitoring Questions

Another method that students can use for self-assessment is to ask themselves a set of questions as a check on their engagement in self-awareness and self-evaluation during mathematics class. In looking at the problem-solving activities of high school students, Schoenfeld (1985) suggested a set of self-monitoring questions, shown in figure 26.2, that were aimed at encouraging students to take control of their own learning. The first question evokes self-awareness by asking students to describe their activities, but it is not meant to prevent exploration. An answer like "I am trying out a few numerical examples to see if I understand the problem" has a much more positive connotation than an answer like "I really don't know what I am doing." The two other questions are more self-evaluative in that they invite students to take a critical look at the activities they have chosen. Once again, some answers can show that the

- What (exactly) are you doing?
  (Can you describe it precisely?)
- Why are you doing it?
  (How does it fit into the solution?)
- How does it help you?
  (What will you do with the outcome when you obtain it?)

Fig. 26.2. Self-monitoring questions (from Schoenfeld [1985, p. 374])

student is in control of the situation (*I'm making the list to look for a pattern*), whereas other answers show that the student is floundering (*I really don't know why I am doing this*).

As they become accustomed to using self-monitoring questions as a catalyst for self-assessment, students can internalize the questions, provide better answers to them, and modify them to suit personal needs and styles. Regardless of what questions are used and how students provide answers to their questions, this method of self-assessment serves to encourage self-awareness and self-evaluation in students and ensure that they are "in control" of their learning.

Internalizing Criteria for Judgments

Students at all grade levels are notorious for asking, "Is this right?" One reason for students to become adept at self-assessment is to equip themselves with the ability to answer that question for themselves and answer it in ways that are consistent with the kind of judgments that teachers themselves might make. Techniques such as completing charts or thinking about answers to appropriate self-monitoring questions can prove to be fruitless endeavors unless students have also internalized appropriate criteria for judging the quality of their own work.

Sources of benchmarks and criteria that students can use to evaluate their own work and establish their own set of judgment criteria appear in reports on alternative mathematics assessment (e.g., California State Department of Education 1989; Pandey 1991; Mumme 1990; Stenmark 1989, 1991; Vermont Department of Education 1991). Scoring guides and rubrics, such as the California Assessment Program's performance standards for student work that appear in figure 26.3, become models that students can use in a variety of self-assessment activities. For example, one activity involves giving a scoring rubric and a set of scored responses to a group of students and having them work collaboratively to explain why a particular response received a particular score. By establishing the match between the score and the scoring criteria, students have the opportunity to internalize the criteria that were used in scoring the tasks. As an extension exercise, the students could create new examples of responses appropriate for each of the score levels. A more "constructivist" version of this activity would involve students studying a set of unscored student responses and developing their own scoring rubric based on their own set of evaluative criteria.

This use of scoring rubrics and sample responses may require some modification of printed materials before they can be used with precollege mathematics students, since materials such as content requirements and scoring guides are generally written for a relatively sophisticated adult audience. Also, the usefulness of the mathematics tasks will be limited by the extent to which they are aligned with the local mathematics curriculum. If possible, it would be more meaningful for students to use locally produced open-ended mathematics

Level	Standard to be achieved for performance at specified level
6	Fully achieves the purpose of the task, while insightfully interpreting, extending beyond the task, or raising provocative questions. • Demonstrates an in-depth understanding of concepts and content. • Communicates effectively and clearly to various audiences, using dynamic and diverse means.
5	Accomplishes the purposes of the task. • Shows clear understanding of concepts. • Communicates effectively.
4	Substantially completes purposes of the task. • Displays understanding of major concepts, even though some less important ideas may be missing. • Communicates successfully.
3	Purpose of the task not fully achieved; needs elaboration; some strategies may be ineffectual of not appropriate; assumptions about the purposes may be flawed. • Gaps in conceptual understanding are evident. • Limits communication to some important ideas; results may be incomplete or not clearly presented.
2	Important purposes of the task not achieved; work may need redirection; approach to task may lead away from its completion. • Presents fragmented understanding of concepts; results may be incomplete or arguments may be weak. • Attempts communication.
1	Purposes of the task not accomplished. • Shows little evidence of appropriate reasoning. • Does not successfully communicate relevant ideas; presents extraneous information.

Fig. 26.3. Performance standards for student work (from Pandey 1991, p. 30)

tasks, sample responses, and scoring guides (or even their own prior successful and not-so-successful experiences) as benchmarks against which they could compare their work and establish criteria for judgment. Whatever the source, however, scoring rubrics and sample responses from alternative mathematics assessments present yet another way in which students can engage in self-assessment.

THE TEACHER'S ROLE IN STUDENT SELF-ASSESSMENT

Ideally, students should be initiating and orchestrating self-assessment activities for themselves. However, it is generally accepted that students, especially those in the elementary grades, are not always adept at describing or judging their own abilities (Brown 1978) and that students at all grade levels often repeatedly fail to use self-monitoring and self-evaluative strategies (Schoenfeld 1985). How, then, can students learn to become successful self-assessors in mathematics? One answer to this question is through the intervention of classroom mathematics teachers.

Teachers can provide experiences through which students can develop their own capability for self-assessment by choosing or designing instruments to which students would respond. The instruments can foster self-assessment in areas such as mathematical knowledge, processes, and attitudes; the response formats can be simple checklists (yes-no, true-false), response scales (Strongly Agree to Strongly Disagree), sentence starters (Today in mathematics I learned ...), or focus questions (How did you feel about this group activity?) Student self-assessment instruments for mathematics can be found in a variety of sources (e.g., Charles, Lester, and O'Daffer 1987; Clarke 1988; Stenmark 1989, 1991).

However, responding to a set of predetermined questions or statements may not necessarily guarantee that students are actually engaged in self-assessment. Nevertheless, there are some fairly simple ways in which even elementary school students can be encouraged to be more reflective about their own learning and performance in mathematics. After they had finished selected lessons in a unit, twelve- and thirteen-year-old students were asked to write responses to the following two questions:

- What did you learn today? Explain the main point(s) of today's lesson.
- On what points are you still confused? Explain.

(This example was adapted from a similar activity observed by the second author during his fall 1988 visits to elementary and lower secondary schools in Japan. Some additional aspects of those classroom visits and the observed mathematics teaching are summarized in Becker et al. [1990].) The teacher used the responses to provide feedback on the effectiveness of his communication of the major theme of the lesson and to help him plan future instruction. But he frequently summarized the responses and shared the summary with the class. The students worked in pairs to discuss their own responses to the two questions and then participated in a class discussion about their different views of the main points.

By giving his students summaries of the class responses and encouraging discussion, this teacher presented his students with an opportunity to record their impressions of the mathematics lesson (self-awareness) and then to reflect on their own responses (self-evaluation), processes in which they may not have engaged without being prompted. In addition, this kind of activity also benefited the teacher by adding to what he knew about his students and by furnishing information that could influence instruction.

Another way to encourage student self-assessment is for teachers to use simple classroom exercises that can be easily incorporated into normal classroom activities at opportune times. Like the self-assessment instruments, these activities can encourage self-awareness and self-evaluation. Also, they serve to blur the edges between self-assessment and mathematics instruction.

CONCLUSION

One goal of a mathematics program based on the NCTM *Curriculum and Evaluation Standards for School Mathematics* (1989) and the *Professional Standards for Teaching Mathematics* (1991) is for students to gain mathematical power. One important attribute of mathematically powerful learners is their ability to know how much they know, to judge the quality of this knowledge, and to know what they need to do in order to learn more. These characteristics are also at the heart of student self-assessment. For students, self-assessment encourages them to assume an active role in the development of mathematical power. For teachers, student self-assessment activities can provide a lens through which the development of students' mathematical power can be viewed.

REFERENCES

Becker, Jerry P., Edward A. Silver, Mary Grace Kantowski, Kenneth J. Travers, and James Wilson. "Some Observations of Mathematics Teaching in Japanese Elementary and Junior High Schools." *Arithmetic Teacher* 38 (October 1990): 12–21.

Brown, Ann L. "Knowing Where, When, and How to Remember: A Problem of Metacognition." In *Advances in Instructional Psychology,* Vol. 1, edited by Robert Glaser, pp. 77–165. Hillsdale, N.J.: Lawrence Erlbaum Associates, 1978.

Brown, Ann L., John D. Bransford, Roberta A. Ferrara, and Joseph C. Campione. "Learning, Remembering, and Understanding." In *Handbook of Child Psychology,* Vol. 3, edited by John H. Flavell and Ellen M. Markman, pp. 77–166. New York: John Wiley & Sons, 1983.

California State Department of Education. *A Question of Thinking.* Sacramento, Calif.: The Department, 1989.

Campione, Joseph C., Ann L. Brown, and Michael L. Connell. "Metacognition: On the Importance of Understanding What You Are Doing." In *The Teaching and Assessing of Mathematical Problem Solving,* edited by Randall I. Charles and Edward A. Silver, pp. 93–114. Reston, Va.: National Council of Teachers of Mathematics, 1988.

Charles, Randall, Frank Lester, and Phares O'Daffer. *How to Evaluate Progress in Problem Solving.* Reston, Va.: National Council of Teachers of Mathematics, 1987.

Clarke, David. *Assessment Alternatives in Mathematics.* Canberra, Australia: Curriculum Development Centre, 1988.

Flavell, John H. "Metacognitive Aspects of Problem Solving." In *The Nature of Intelligence,* edited by Lauren B. Resnick, pp. 231–35. Hillsdale, N.J.: Lawrence Erlbaum Associates, 1976.

_____. "Metacognition and Cognitive Monitoring: A New Area of Cognitive-Developmental Inquiry." *American Psychologist* 34 (1979): 906–11.

Garofalo, Joe, and Frank K. Lester, Jr. "Metacognition, Cognitive Monitoring, and Mathematical Performance." *Journal for Research in Mathematics Education* 16 (May 1985): 163–76.

Krutetskii, Vadim A. *The Psychology of Mathematical Abilities in School Children.* Translated by Joan Teller. Chicago: University of Chicago Press, 1976.

Lester, Frank K. "Reflections about Mathematical Problem-solving Research." In *The Teaching and Assessing of Mathematical Problem Solving,* edited by Randall I. Charles and Edward A. Silver, pp. 115–24. Reston, Va.: National Council of Teachers of Mathematics, 1988.

Mumme, Judy. *Portfolio Assessment in Mathematics.* Santa Barbara, Calif.: California Mathematics Project, University of California, Santa Barbara, 1990.

National Council of Teachers of Mathematics. *Curriculum and Evaluation Standards for School Mathematics.* Reston, Va.: The Council, 1989.

_____. *Professional Standards for Teaching Mathematics.* Reston, Va.: The Council, 1991.

Pandey, Tej. *A Sampler of Mathematics Assessment.* Sacramento, Calif.: California Department of Education, 1991.

Resnick, Lauren B. "Treating Mathematics as an Ill-structured Discipline." In *The Teaching and Assessing of Mathematical Problem Solving,* edited by Randall I. Charles and Edward A. Silver, pp. 32–60. Hillsdale, N.J.: Lawrence Erlbaum Associates, 1989.

Schoenfeld, Alan H. "Metacognitive and Epistemological Issues in Mathematical Understanding." In *Teaching and Learning Mathematical Problem Solving: Multiple Research Perspectives,* edited by Edward A. Silver, pp. 361–79. Hillsdale, N.J.: Lawrence Erlbaum Associates, 1985.

_____. "What's All the Fuss about Metacognition." In *Cognitive Science in Mathematics Education,* edited by Alan H. Schoenfeld, pp. 189–216. Hillsdale, N.J.: Lawrence Erlbaum Associates, 1987.

Silver, Edward A. "Research on Teaching Mathematical Problem Solving: Some Underrepresented Themes and Needed Directions." In *Teaching and Learning Mathematical Problem Solving: Multiple Research Perspectives,* edited by Edward A. Silver, pp. 247–66. Hillsdale, N.J.: Lawrence Erlbaum Associates, 1985.

Silver, Edward A., Nicholas Branca, and Verna Adams. "Metacognition: The Missing Link in Problem Solving?" In *Proceedings of the Fourth International Conference for the Psychology of Mathematics Education,* edited by Robert Karplus, pp. 213–21. Berkeley, Calif.: University of California, 1980.

Silver, Edward A., Jeremy Kilpatrick, and Beth Schlesinger. *Thinking through Mathematics.* New York: College Entrance Examination Board, 1990.

Stenmark, Jean Kerr. *Assessment Alternatives in Mathematics.* Berkeley, Calif.: EQUALS, University of California, 1989.

_____. *Mathematics Assessment: Myths, Models, Good Questions, and Practical Suggestions.* Reston, Va.: National Council of Teachers of Mathematics, 1991.

Vermont Department of Education. *Looking beyond "The Answer": The Report of Vermont's Mathematics Portfolio Assessment Program.* Montpelier, Vt.: The Department, 1991.

27

Assessment, Understanding Mathematics, and Distinguishing Visions from Mirages

Thomas J. Cooney
Elizabeth Badger
Melvin R. Wilson

MATHEMATICS education today is the subject of intense pressure for reform. Part of the pressure is focused on the importance of evaluation at all levels of schooling. This is quite appropriate, for evaluation can be construed not only as an object of reform but as an instrument of reform as well. The means by which we assess our students' understanding of mathematics have the potential for helping us realize the vision set forth in the *Curriculum and Evaluation Standards for School Mathematics* (NCTM 1989).

THE IMPORTANCE AND MEANING OF ASSESSMENT

It is imperative that assessment be seen as an integral part of instruction. It provides a window to students' thinking and a compass for instruction. It helps teachers maintain a perspective of reform in the face of numerous obstacles. Equally important, what gets assessed—and how it gets assessed—sends clear signals to students about what teachers believe is important. As Rogers (1969, p. 956) suggested,

> Examinations tell them our real aims, at least so they believe. If we stress clear understanding and aim at a growing knowledge of physics, we may completely sabotage our teaching by a final examination that asks for numbers to be put into memorized formulas. However loud our sermons, however intriguing the experiments, students will judge by that examination—and so will next year's students who hear about it.

The literature is filled with instances in which students view as suspect any content or teaching method that they perceive as deviating from what they consider the norm. Borasi (1990) refers to this circumstance as the "invisible hand" that influences instruction. This influence is usually narrow in focus and predictable and stems from a limited conception of what it means to know and do mathematics. She identifies four beliefs of students about mathematics that contribute to a classroom environment best characterized as limited or "dualistic" (p. 176):

- The scope of mathematical activity consists primarily of identifying a correct answer to a well-defined problem that has an exact and predetermined solution.

- The nature of mathematical activity consists primarily of recalling and applying learned procedures to solve problems.

- Mathematical knowledge consists of facts and procedures that are fixed, predetermined, and not subject to human judgment.

- Mathematical knowledge exists as a finished product; the task of the mathematician is to discover that knowledge.

TEACHERS' CONCEPTIONS OF UNDERSTANDING MATHEMATICS

During the summer of 1990 and the subsequent school year, the authors embarked on a project to study teachers' assessment practices at the middle and secondary school level in order to understand how teachers assess students and why teachers value what they do. (See Cooney [1992].) The study involved 201 middle and secondary school mathematics teachers who completed an initial questionnaire that focused on general assessment practices, 102 of these teachers who completed a subsequent questionnaire in which they reacted to five different types of assessment items, and a detailed interview with 18 of the teachers who further elaborated on their means of assessing mathematical understanding. Although the study had many foci, we shall concentrate here on how the teachers assessed their students' understanding of mathematics.

Teachers' Interpretation of Understanding Mathematics

Nearly one-half (48%) of the teachers who completed the initial questionnaire indicated that the primary source of their unit or chapter tests was textbook publishers. Smaller percentages of teachers indicated that they created their own tests or used other teachers' tests. This immediately raises a question concerning teachers' confidence in their ability to create the kinds of questions that can assess their students' understanding of mathematics on a broad range of indicators. More specifically, what interpretations do teachers give to the notion of assessing a "deep and thorough understanding" of mathematics? To investigate this issue, we posed the following two tasks with reference to one of three topics—fractions, area, or functions (linear, quadratic, other):

1. Write or draw a typical problem that you gave students that you believe tests a minimal understanding of the topic.

2. Write or draw a typical problem that you gave students that you believe tests a deep and thorough understanding of the topic.

All responses were categorized among the following four "levels":

Level 1: A problem calling for simple computation or recognition

Level 2: A problem requiring comprehension (The student must make some decision, but once made, the solution process is straightforward, e.g., a simple one-step word problem.)

Level 3: An application or multistep problem in which the student must make several decisions about process or what operations to use

Level 4: A nonroutine or open-ended problem

Typical responses are given in figure 27.1.

We found that a majority (56%) of the 201 teachers created an item at Level 1 or 2 to test a deep and thorough understanding of mathematics. Among this group was a very high percentage (87%) of teachers who generated problems related to fractions. Perhaps this topic entices teachers to use lower-level items

Topic	Level 1	Level 2	Level 3	Level 4
Area	Find the area of a rectangle with a width of 4 inches and a length of 2 inches.	Find the area of the parallelogram $5 \diagup 4$ over 8	Find the surface area: $5 \diagdown 3$ over 4	Draw the floor plan of a house and determine the number of square feet in the house.
Functions	1. Is the relation{(0,5), (1,3), (0,7), (2,4), (3,9)} a function? 2. Graph $y = 2x + 3$.	1. Find the equation of the line containing the points (2,3) and (−1,5). 2. Graph $y = (x − 3)^2 − 2$	Find the equation of the line through (−2,3) and perpendicular to the line $2y + 5x = 5$.	Write a quadratic function $f(x)$. Write the function that would translate $f(x)$ vertically; horizontally; dilate f.
Fractions	1. $\dfrac{3}{8} + \dfrac{1}{4} = ?$ 2. What fraction of the rectangle is shaded?	Mary and Joe are taking a trip of 80 miles. Mary drove 2/5 of the distance. How many miles did she drive?	Bob ate 1/4 of a pepperoni pizza, 2/3 of a cheese pizza, and 1/2 of a sausage pizza. How much pizza did he eat?	Identify the activities of a typical teenager in a 24-hour period. Graphically represent the fractional parts of a day spent on these activities.

Fig. 27.1. Typical problems generated by teachers

to assess a deep understanding of mathematics. However, since teachers who selected fractions also tended to be middle school teachers or teachers of low-ability students, an alternative explanation may be that these teachers did not believe that a broad repertoire of indicators was appropriate for their students. In fact, 73 percent of the thirty-three teachers of below-average students created only Level 1 or 2 items. A very typical Level 1 item to test a deep and thorough understanding of fractions was to add or subtract more complicated fractions—for example,

$$4\frac{1}{5} + 3\frac{2}{3}.$$

The subsequent interviews gave further insight into teachers' conceptions of mathematics and assessment. During these interviews, teachers used various terms to express the notion of determining whether their students held a "deep and thorough" understanding of mathematics. Two teachers talked about having students visualize processes—for example, "changing equations into graphs." Another teacher emphasized the notion of synthesis—creating items that drew upon several different mathematical domains. Yet another teacher discussed the importance of students exhibiting their reasoning processes as they used different algorithms.

Most of the teachers, however, used the language of "steps" in discussing the teaching and learning of mathematics and in identifying or creating the kind of items needed to assess a deeper understanding of mathematics. The following quotes were typical:

- It takes two-part problems, not just one-step problems [to test a deep and thorough understanding].
- [A deep and thorough understanding] involves multiple steps.
- I would at least want the child to show some of his steps.
- [A deep and thorough understanding] is the ability to break a complex problem into simple steps.
- Mathematics needs to be broken down into small steps.
- I teach step by step.

Solving multistep problems is not an altogether inappropriate means of assessing a deeper understanding of mathematics. Certainly the ability to factor the expression $x^4 - 1$ is more indicative of an understanding of factoring the difference of two squares than the ability to factor the expression $x^2 - 1$. Nevertheless, the notion that the complexity of mathematics is based on the accumulation of simpler steps fails to capture the notion of mathematical power suggested in the *Standards* (NCTM 1989).

Perceived Usefulness of Performance Items

In order to investigate the extent to which teachers were prepared to use more integrated items, we posed in our second questionnaire five problems that, to varying degrees, involved communication, reasoning, and problem solving. Teachers were asked to solve each problem; to identify the mathematical content and processes tested; and to indicate the likelihood of their using each problem, as well as the context in which it might be used.

The choice of problems was based on the content and processes each problem measured. Figure 27.2 illustrates the five problems in the order they appeared in the questionnaire. The first problem required the interpretation of a functional relationship; two problems (items 2 and 3) focused on students' understanding of the concept of area; two others (items 4 and 5) involved the use of fractions. Although each problem was set within the context of a familiar topic, as a whole the problems went beyond the content itself to measure students' ability to reason, communicate, or apply their knowledge. In other words, the problems called for an understanding that we would characterize as "deep and thorough."

Teachers' answers for each of the five problems were coded as well as the likelihood that they would use each question when assessing their students. Four of the five problems were coded holistically (ranges are given in table 27.1); the remaining question was scored 0 or 1. Mean "correctness" scores (Avg./Maximum) were obtained by dividing the average score by the maximum possible score. The likelihood scores ranged from 1 to 5, 1 being "very likely" to use the item, 5 being "very unlikely" to use the item. Table 27.1 illustrates what percent of the teachers indicated they were likely or very likely to use the various items (score of 1 or 2).

Table 27.1
Summary of Teachers' Reactions to Performance Items

(N = 101)	Item 1	Item 2	Item 3	Item 4	Item 5
Possible Scores	0–3	0–3	0–1	0–3	0–2
Average Score	2.32	2.31	.95	1.80	1.97
Avg./Maximum	.77	.77	.95	.60	.98
Likely or Very Likely to Use	54%	66%	79%	75%	85%

Teachers experienced the most difficulty with the fraction problem (item 4) and were most successful in answering the area and picture problems (items 3 and 5). This may be accounted for by the fact that the area and picture problems call for the identification of a specific number, but the other problems require some sort of explanation. This finding has implications for instruction. If

1. A researcher asked many students two questions: "What was your grade on your last math exam?" and "How many hours per night did you usually spend on math homework?" She then sorted students into groups according to how much time they had spent. Finally, she computed an average grade for each group and plotted the averages in the graph. Write a plausible explanation to explain the data.

Grade in Mathematics
(Numerical Grade)

Time Spent on Mathematics
Homework Each Evening
(Average)

2. Theo wants to find out which pond covers the larger area, Parker Pond or Shelby Pond. He does not need to know the two areas, just which is bigger. Theo claims that all he has to do is measure the distance around each pond to find out what he wants.

Will Theo's method work? Write a convincing argument for your answer.

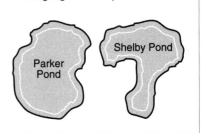

Shelby Pond

Parker Pond

3.

⟨ • ⟩ is one unit of area.

Given the unit of area shown above, what is the area of the larger figure?_____

4. Gwen was given the problem 2/5 < ? < 4/7. She said that 3/6 would be between 2/5 and 4/7. The teacher asked Gwen to explain how she got her answer and why she thinks her method works. Gwen said that she chose a numerator of 3 because 2 < 3 < 4 and a denominator of 6 because 5 < 6 < 7. Gwen claimed her method always works and gave the following examples:

i. The fraction 2/4 is between 1/3 and 3/5 because 1< 2 < 3 and 3 < 4 < 5.

ii. The fraction 4/9 is between 2/5 and 6/11 because 2 < 4 < 6 and 5 < 9 < 11.

Does Gwen's method always work? Explain your reasoning.

5.

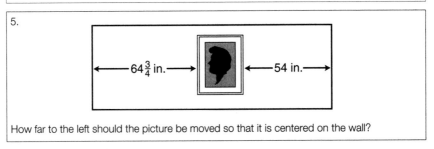

←— $64\frac{3}{4}$ in. —→   ←—54 in.—→

How far to the left should the picture be moved so that it is centered on the wall?

Fig. 27.2. Items from Phase II of survey. *Note:* Items 1, 2, 3, and 5 were developed as part of the Massachusetts Assessment Program; item 4 was adapted from the *Standards* (NCTM 1989, p. 230).

teachers find it difficult to answer items that require the construction of counterexamples or arguments, then it seems that they will not be likely to engage their students in such activities. The fact that the teachers indicated that they would more likely use the area or picture problems lends credence to the notion that they favor problems whose solutions consist of a single number.

Teachers gave a variety of reasons for their reluctance to use the items, including a lack of confidence in answering such questions, the view that the items were too difficult or not appropriate for a testing situation, and the feeling that the items did not test appropriate mathematics for their students. Other issues that the teachers raised during the interviews included concern over increased demands for time, the issue of fairness in grading, and the fact that few resource materials were available to help them use such items. The issue of fairness was expressed in several ways, for example, concerns about scoring items holistically and the notion that the item might be appropriate for class discussion when assessments and judgments are not required.

What is the relevance of these findings to classroom practice? Are open-ended questions that explore deep understanding of mathematical processes irrelevant to the practical concerns of everyday classroom life—the grading charts, the textbook exercises, the need for students to acquire the facts and skills that form the basis for mathematical thinking? Will students see the relationship between open-ended, challenging questions and their understanding of mathematics? Is the whole idea of "authentic assessment" alien to what goes on in the classroom? We think not, or at least it need not be. In the following sections, we shall explore how a broader conception of assessment can be compatible with the reality of the classroom.

ALTERNATIVE ASSESSMENT TASKS

Tasks that provide insight into a student's understanding of mathematics are often of a different form from ones typically found on tests. They possess the following characteristics:

- A good assessment task involves significant mathematics.

"Where is the time to do this?" is not a trivial question. Good tasks require time and thought on the part of teachers and students. Beyond this, tasks used in assessment carry a message of what is important in mathematics for students, their parents, and the public.

- A good assessment task can be solved in a variety of ways.

Too often in mathematics classrooms the message is communicated that there is only one correct way to complete a procedure or solve a problem. This is seldom true when problems are solved in the real world. To encourage confidence, flexibility, creativity, and persistence, assessment problems should hold the possibility for alternative solutions, and teachers should be ready to accept solution methods that are not the ones that they themselves had in mind.

- A good assessment task elicits a range of responses.

Concepts are not like light switches to be turned on or off; one doesn't lack a concept one moment and have it the next. Concepts are built slowly, growing in complexity over a lifetime of experiences. For this reason, tasks should be open enough to be accessible to all students—from those whose understanding is limited to those for whom the concepts have become powerful vehicles for mathematical thinking.

- A good assessment task requires students to communicate.

In and of itself, a correct answer gives very limited information. It does not even signify whether the solution process was correct. Students' writing, drawing, and symbolizing, on the other hand, illuminate their thinking. Teachers gain insight into how students view mathematics as well as their ability to apply it.

- A good assessment task stimulates the best possible performance on the part of the student.

For many students, the typical mathematical "problem" is a challenge in recognition and memory. It seldom taps students' knowledge in contexts beyond the classroom walls. In contrast, tasks that are relevant to students' lives or that contribute to a student's understanding not only assess their mathematical ability but also convey the fascination and utility of mathematics. If assessment tasks are a culmination of learning, they should pose problems that require students to synthesize or apply what they have learned.

DECIDING WHAT IS TO BE ASSESSED

Knowledge, vocabulary, number facts, formulas, and the attributes of figures form the knowledge base of mathematical activity. They are acquired through practice and represent the kind of mathematics that can easily be seen as "correct/incorrect," "right/wrong." It would be foolish to ignore such mathematical outcomes. Although we do not want students to perceive these topics as the essence of "real mathematics," ignoring them not only does students a disservice in terms of their own learning but robs them of an area in which most can be successful.

But there is another important component of mathematical knowledge where a right/wrong judgment has little relevance. These are the "big ideas" of mathematics, ideas that continue through a lifetime of growth and elaboration. The hallmark for difficulty in this area is not the complexity of the data but the complexity of ideas. How students understand a particular concept tells us much about their mathematical understanding in general. This type of knowledge can be judged only by a response that requires a deeper understanding of mathematics.

DECIDING HOW TO ASSESS

Developing good tasks is not an easy enterprise. Furthermore, despite the most prolific imagination and herculean efforts, few will manage to produce a task that fulfills all the criteria set forth above. A more profitable route may be to adapt more easily available "traditional" items in order to better assess students' mathematical understanding. Below are some examples of the kinds of transformations that can be used:

1. Placing ordinary items in a practical context
2. Asking for the reasons behind procedures
3. Encouraging critical analysis

BRIDGING THE GAP

Unless it is understood that new forms of assessment reflect a different and more fundamental vision of what it means to know mathematics, rich and challenging tasks not only will have little relevance to the curriculum but will not be used. Our research has indicated that however innovative the tasks, teachers will not use them for assessment if (1) these tasks do not reflect their own understanding of mathematics, (2) they do not recognize the value of such tasks in measuring significant mathematical knowledge, and (3) they do not value the outcomes the items purport to measure. As long as mathematics is viewed as consisting primarily of a series of steps to be applied in isolated contexts, teachers will view moves toward alternative methods of assessment as peripheral to the "real curriculum." As a consequence, a promising vision evaporates into a mirage.

REFERENCES

Borasi, Raffaella. "The Invisible Hand Operating in Mathematics Instruction: Students' Conceptions and Expectations." In *Teaching and Learning Mathematics in the 1990s,* 1990 Yearbook of the National Council of Teachers of Mathematics, edited by Thomas J. Cooney, pp. 174–82. Reston, Va.: The Council, 1990.

Cooney, Thomas J. *A Survey of Secondary Teachers' Evaluation Practices in Georgia.* A study supported by the Eisenhower Program for the Improvement of Mathematics and Science Education. Athens, Ga.: University of Georgia, 1992.

Rogers, E. M. "Examinations: Powerful Agents for Good or Ill in Teaching." *American Journal of Physics* 37 (1969): 954–62.

National Council of Teachers of Mathematics. *Curriculum and Evaluation Standards for School Mathematics.* Reston, Va.: The Council, 1989.

Index